Comprehensive
Science
Assessment

4

DISCARDED

Options
Publishing

Acknowledgments

Executive Editor: Linda Bullock

Editor: Carolyn Thresher

Production Supervisor: Sandy Batista

Senior Production Specialist: Corrine Scanlon

Product Development: Science House Publishing Services

Design and Production: Think Design Group

Cover Design: Christine Grether

ISBN 1-59137-514-2

Options Publishing
P.O. Box 1749
Merrimack, NH 03054-1749
Toll Free: 800-782-7300
Fax: 866-424-4056
www.optionspublishing.com

Printed in the U.S.A.

15 14 13 12 11 10 9 8 7 6 5 4 3 2 1

Table of Contents

To the Student

Comprehensive Science Assessment helps you prepare for your science tests by

- reviewing important science ideas
- recalling science vocabulary
- testing yourself
- practicing answering science questions

There are four kinds of pages in this book.

Lessons

- Lessons start with a question. The question tells you what you will read about in each lesson.
- Every lesson includes a picture, diagram, graph, or chart. These help you understand what you are reading.
- Each lesson ends with *Show What You Know*. Questions and activities in this part of the lesson help you review important science concepts.

Tests

Each test includes two kinds of questions—multiple-choice and short-response. Taking these tests helps you know what you have learned.

Test Answer Guides

There is a Test Answer Guide for each test in your book. It explains the correct answer for each item on the test.

Practice Test

At the back of your book, you will find a full-length practice test. Just like the other tests in your book, this test includes two kinds of questions—multiple-choice and short-response.

Basic Needs of Living Things

What do organisms need to grow and live?

An **organism** is a living thing. Plants, animals, mushrooms, and people are examples of organisms. All organisms need **energy** and raw materials to stay alive. Raw materials include nutrients, oxygen, and water. **Nutrients** are materials that organisms need to grow and live.

Most of the energy used by living organisms comes from the sun. Plants trap energy from sunlight and use it to make food. **Food** contains energy and nutrients. Most organisms use oxygen to release the energy stored in food.

Plants make their own food. **Photosynthesis** is the process plants use to make food. Sunlight provides the energy needed for photosynthesis. During photosynthesis, carbon dioxide and water are changed into sugar, and oxygen is released into the air as a waste product. Sugar is the plant's food. A plant uses some of the food it makes to live and grow. Extra food may be stored in the plant's roots, stems, and leaves.

Photosynthesis

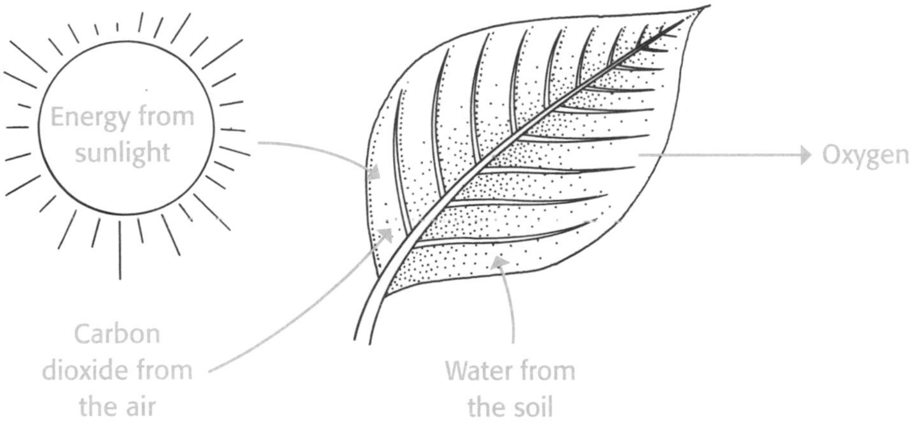

Energy from sunlight

Oxygen

Carbon dioxide from the air

Water from the soil

Show What You Know

Give two reasons why photosynthesis is important to organisms living on Earth.

1. ______________________________

2. ______________________________

LESSON 2

Animal Adaptations

How do body structures help animals meet their needs?

Any characteristic that helps an animal stay alive in its environment is called an **adaptation**. Adaptations help an animal survive to grow up and raise young. Over time, as adaptations are passed from one generation to the next, they increase a species' ability to survive and **reproduce,** or have young.

Body parts are **structural adaptations** that help an animal survive in its environment. Fish use their fins for swimming and their gills for absorbing oxygen from the water. Fins and gills are structural adaptations for living in water.

Some structural adaptations help animals obtain food. Frogs have long, sticky tongues for capturing flies and other insects. Horses have large, flat teeth for chewing grasses.

Camouflage is an adaptation that allows an animal to blend in with its environment. The grayish blue color of a tuna fish matches its underwater surroundings. This camouflage makes it easier for the tuna to sneak up on the smaller fish that are its **prey.** A tuna's camouflage also helps it avoid being seen by a **predator.**

Beak Shapes

A pelican uses its pouch to scoop small fish from the water.

Hummingbirds have long, slender beaks and tongues for reaching deep into flowers to feed on nectar.

Cardinals have stout, cone-shaped bills that can crack hard seeds.

Show What You Know

List one way in which each adaptation helps the animal stay alive.

1. the thick fur of an Arctic wolf ______________________________

2. the powerful jaws of a snapping turtle ______________________________

3. the sharp eyes of a hawk ______________________________

Animal Behaviors

How do behaviors help animals meet their needs?

Behavior is the way an animal acts when it responds to a stimulus. A **stimulus** is anything that makes an animal respond, such as sound, odor, light, heat, hunger, or thirst. A **response** is what an animal does when it senses a stimulus. Thirst is a stimulus. Drinking water is a response.

Many animals change their behavior depending on the season. During the winter, some animals go into a deep sleeplike state called **hibernation.** Hibernation helps the animal survive low temperatures and lack of food. During a hot, dry summer, some animals that require moisture go into another sleeplike state called **estivation**.

A **learned adaptation** is taught to the young by the parents. Kittens, birds, and wolves learn how to hunt for food from their parents. For people and some animals, learned adaptations include using tools.

An **inherited adaptation** is a behavioral adaptation that does not have to be learned. Hibernation, estivation, and migration are inherited adaptations. **Migration** is the movement of animals from one place to another according to the season.

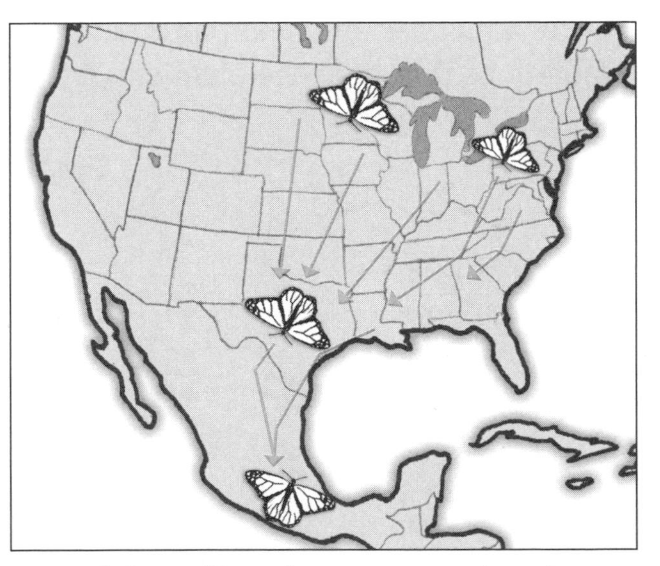

Monarch butterflies migrate to central Mexico for the winter.

Show What You Know

Classify the following human adaptations as either learned or inherited.

1. ____________________ Planting seeds to grow food

2. ____________________ Reading and writing

3. ____________________ A baby grasping a toy with its thumb and fingers

LESSON
4

Plant Adaptations

How do the structures of a plant help it meet its needs?

Plants cover almost every part of Earth's surface and live in many different kinds of environments. Most plants live on land, though there are many that live in water or even on other plants. Plants have the same basic needs as all other organisms—air, water, sunlight, and nutrients.

The structures of a plant are adaptations that help the plant survive and reproduce in its environment. The stem of a cactus is thick and spiny to store water and protect the cactus from being eaten by animals. A vine can grow up the trunk of a tall tree to find the sunlight it needs for photosynthesis. The sweet nectar of an orchid flower attracts insects and birds that help **pollinate** the plant.

Plant Structures

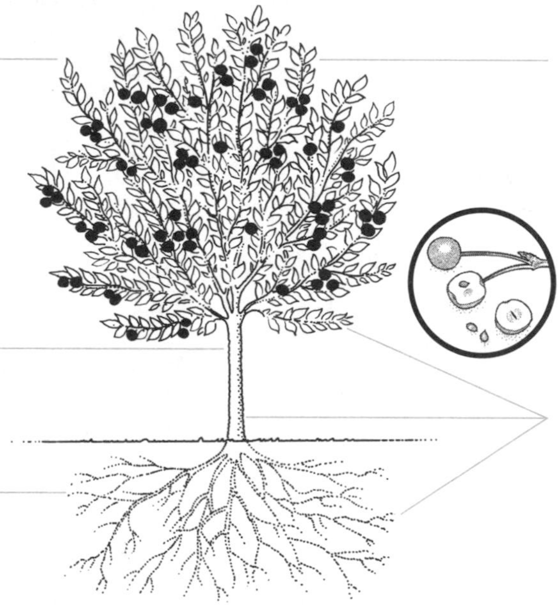

Leaves make food for the plant through photosynthesis.

Many plants produce flowers that develop into fruits with seeds inside.

Most plants produce seeds that can grow into new plants.

The stem holds the plant upright.

Roots take in water and nutrients from the soil. Roots also work like an anchor, keeping the plant in the soil.

Tube-like vessels in roots, stems, and leaves carry water, nutrients, and food to all parts of the plant.

Show What You Know

1. List the basic needs of plants.

______________, ______________, ______________, ______________

2. Name two structural adaptations that help a tree meet its need for water.

______________________, ______________________

Plant Behaviors

How do behaviors help plants meet their needs?

Plant behaviors are adaptations that help plants meet their basic needs.

Response to Light Plants make food through photosynthesis, which requires light. Photosynthesis most often takes place in the leaves. As a plant grows, its stems and leaves turn toward the sunlight.

Response to Gravity The roots of a plant grow down into the soil, where they take in water and nutrients. Roots grow toward the pull of gravity. Stems and leaves grow up, away from the pull of gravity.

Response to Water Like all organisms, plants need water to stay alive. Plants get water through their roots, which grow toward moist places in the soil.

Response to Touch Some plants respond to touch. For example, climbing vines have twisting stems that curl around fences or other plants to find more sunlight.

Show What You Know

Write the plant behavior shown in each diagram. Then explain how each response helps the plant survive in its environment.

1. ______________________________

____

____

2. ______________________________

____

____

3. ______________________________

____

____

LESSON 6

Reproduction

How do animals and plants produce young?

All organisms reproduce. The process of producing **offspring,** or young, is called **reproduction.**

An **inherited trait** is a characteristic a parent passes to its offspring. For example, a baby giraffe inherits a long neck and the ability to digest leaves from its parents. When offspring grow to adulthood, their inherited traits make them similar to their parents. However, offspring are not exactly like their parents. Each offspring inherits different combinations of traits from its parents.

Animals reproduce in many ways. Mammals give birth to live young. Birds lay eggs that have hard shells. Most reptiles, amphibians, fish, and insects also lay eggs. Reptile eggs have leathery coverings. Fish and amphibian eggs are laid in water.

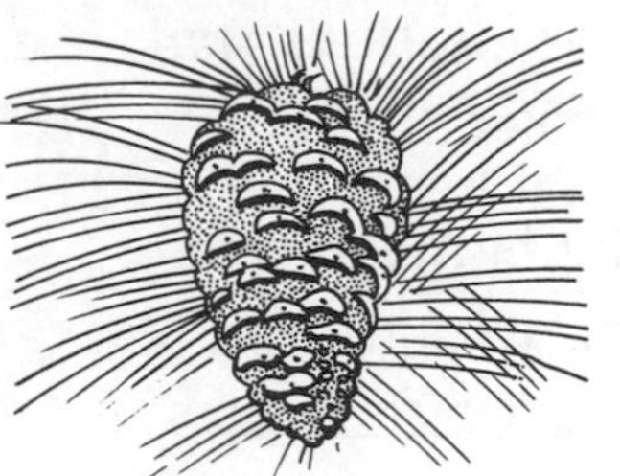

Evergreen trees produce seeds in cones.

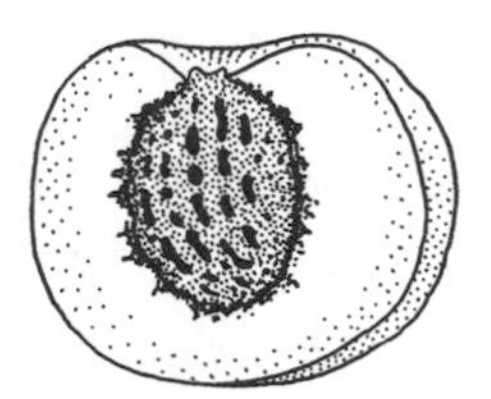

Flowering plants produce seeds inside a fruit.

Plants reproduce by forming seeds or spores that can grow into new plants. **Seeds** are structures that contain a young plant and a food supply. They are protected by a hard outer covering. Seeds are produced in flowers or cones. Ferns and mosses produce spores rather than seeds. **Spores** have a hard outer covering and contain a single cell that can grow into a new plant.

Show What You Know

Complete the chart to show how each organism reproduces.

Organism	How It Reproduces
Mammal	
Flowering plant	
Reptile	
Fish	
Evergreen tree	
Birds	
Amphibians	
Ferns and mosses	

LESSON 7

Animal Life Cycles

How do animals change during their lives?

An animal's **life cycle** is made up of all the stages of **growth** and **development** the animal goes through during its lifetime. Some animals, such as dogs and cats, look like their parents when they are born. They grow larger in size during their lives, but they do not change form. Other animals, including **tadpoles** and **caterpillars**, look nothing like their parents when they are born. These animals change form during their life cycles. The change of form that an animal goes through during its life cycle is called **metamorphosis**.

The Life Cycle of a Frog

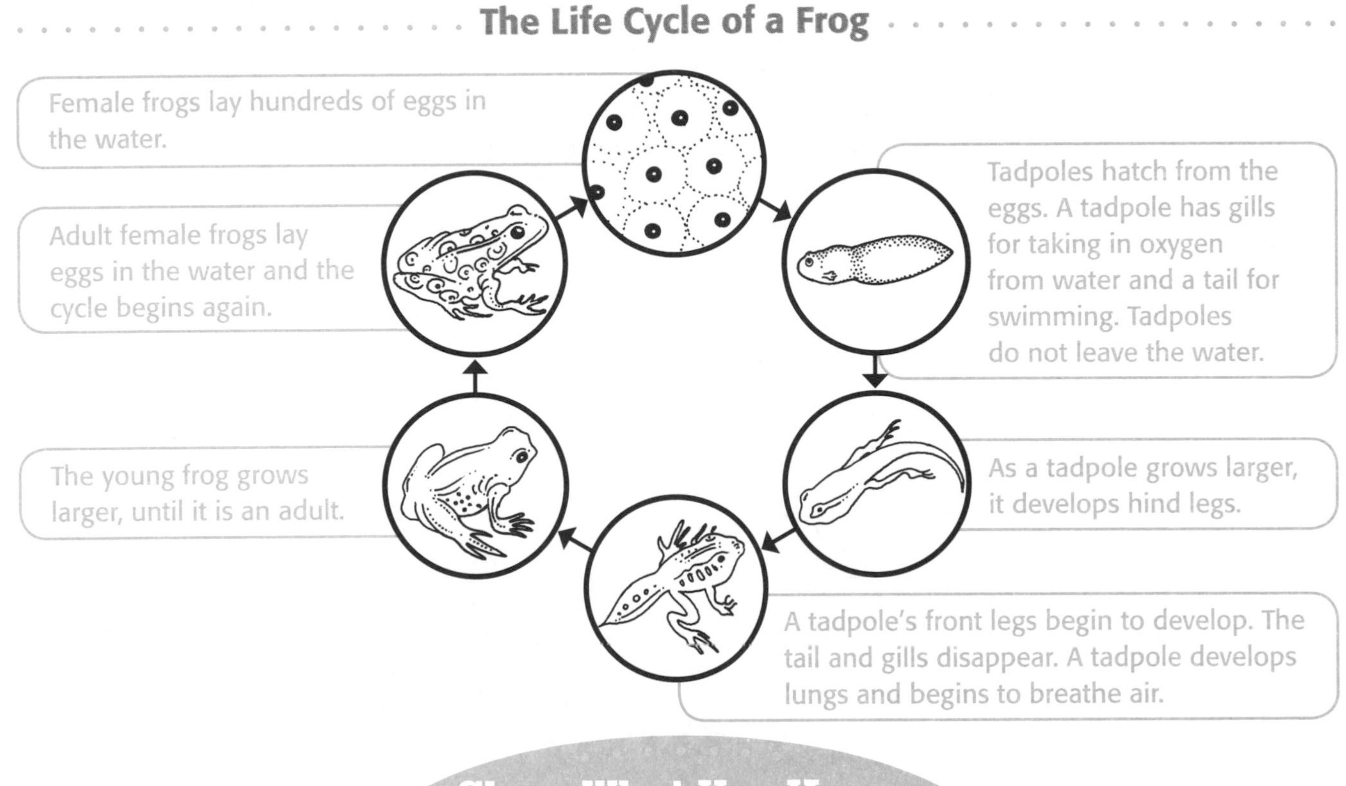

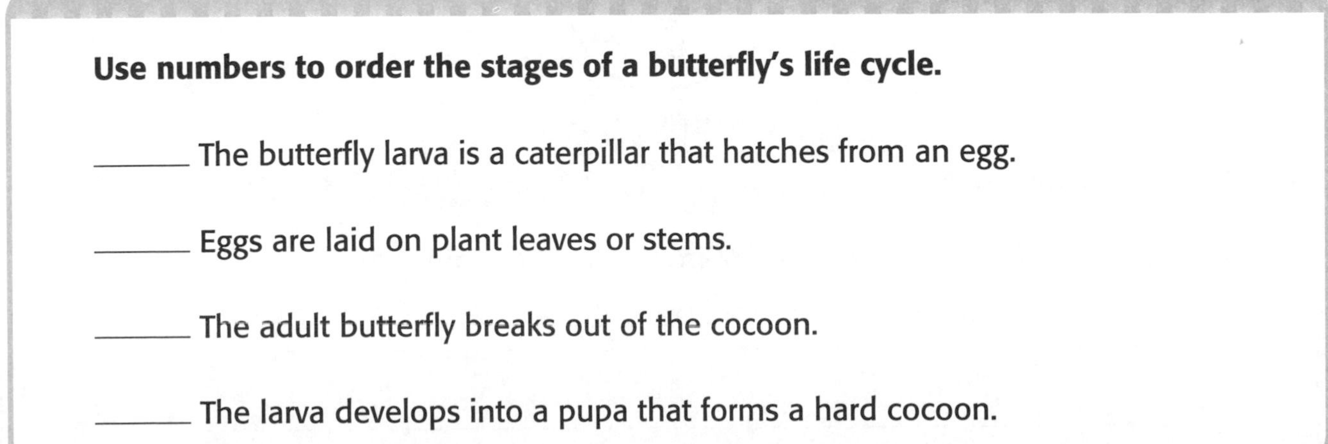

Show What You Know

Use numbers to order the stages of a butterfly's life cycle.

_______ The butterfly larva is a caterpillar that hatches from an egg.

_______ Eggs are laid on plant leaves or stems.

_______ The adult butterfly breaks out of the cocoon.

_______ The larva develops into a pupa that forms a hard cocoon.

LESSON
8

Plant Life Cycles

How do flowering plants change during their lives?

The life cycle of a flowering plant begins with a **seed**. With enough moisture, air, and warmth, a seed will begin to grow, or sprout. This process is called **germination**. The sprouted seed is a **seedling** which can grow into an adult plant that forms flowers.

The **pistil** is the female part of a flower. The **ovary** at the bottom of the pistil contains egg cells that develop into seeds. The **stigma** at the top of the pistil has a sticky surface to catch **pollen.**

The male part of a flower is the **stamen.** The stamen produces pollen. To make seeds, pollen must move from a stamen to a stigma. Usually, wind, insects, birds, and other animals help carry pollen to the stigma of the same or a different flower. This process is called **pollination.**

After pollination, a **fruit** with seeds forms in the ovary. The seeds are released from the fruit and the cycle begins again.

The Life Cycle of a Flowering Plant

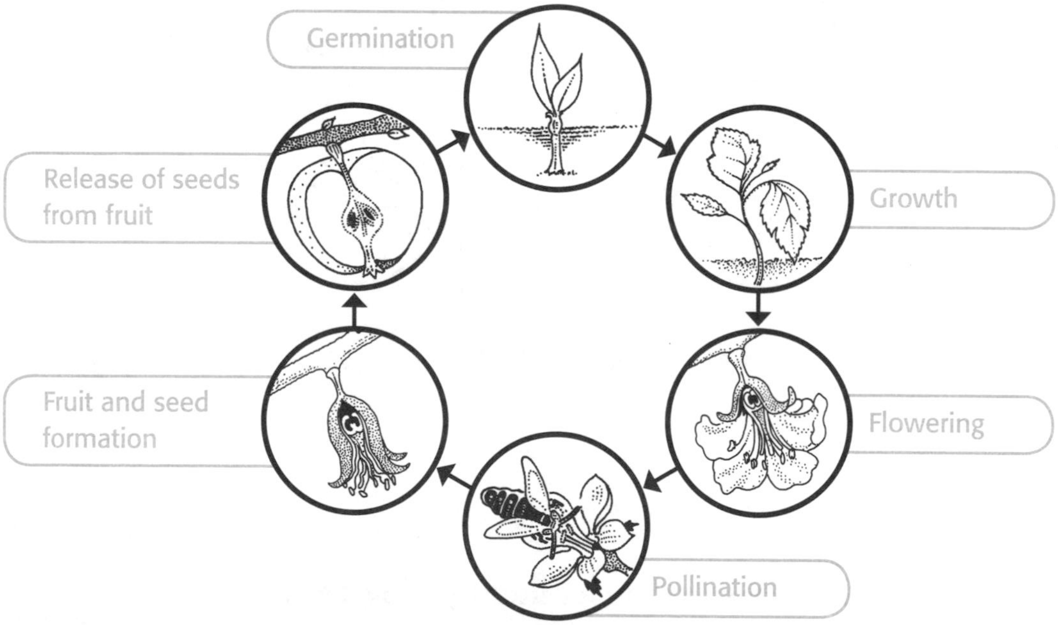

Show What You Know

Explain the difference between pollination and germination.

___

___

Ⓐ Multiple Choice

Fill in the letter to show your answer.

1. **Photosynthesis creates a waste product that is needed by almost all living organisms. This waste product is**

   Ⓐ carbon dioxide.

   Ⓑ oxygen.

   Ⓒ sugar.

   Ⓓ water.

2. **Black bears can spend up to 100 days during the winter without eating or drinking. This kind of adaptation is called**

   Ⓐ estivation.

   Ⓑ camouflage.

   Ⓒ hibernation.

   Ⓓ migration.

3. **Which bird beak would be best adapted for scooping small fish from the water?**

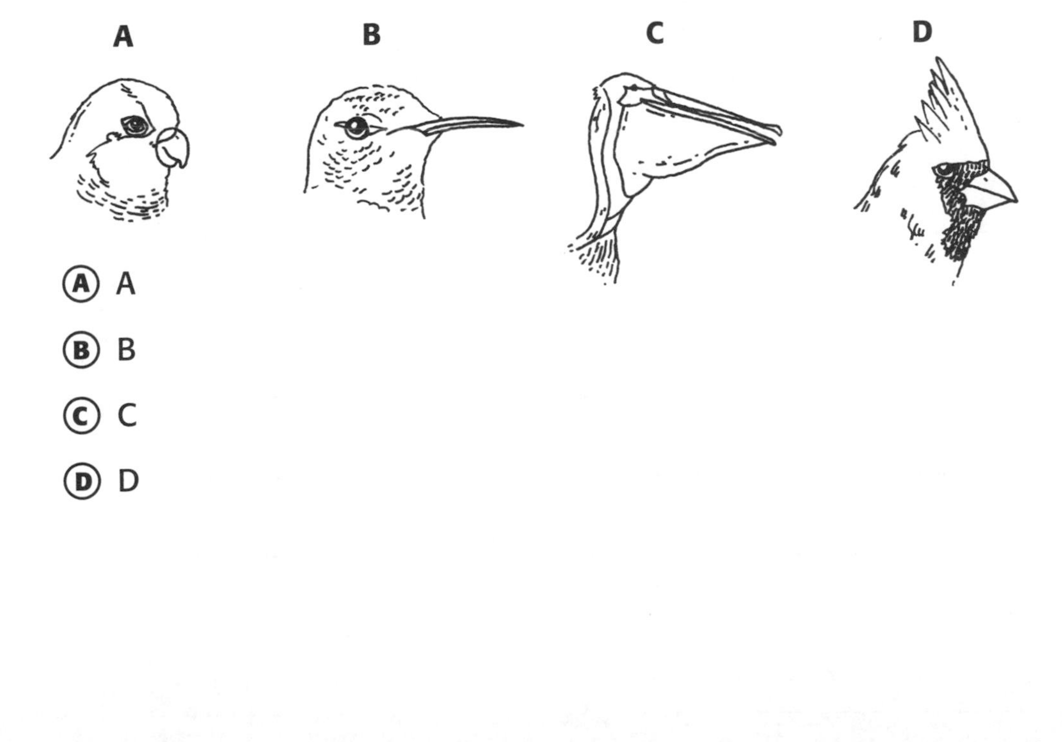

   Ⓐ A

   Ⓑ B

   Ⓒ C

   Ⓓ D

4. **The plant structures in which photosynthesis usually takes place are the**

   Ⓐ leaves.

   Ⓑ stems.

   Ⓒ roots.

   Ⓓ flowers.

5. **Flowering plants reproduce by forming**

   Ⓐ cones.

   Ⓑ spores.

   Ⓒ eggs.

   Ⓓ seeds.

Short Response

6. **Explain how light and gravity affect the growth of plants.**

____

____

____

____

____

____

____

Animal Cells

What's inside an animal cell?

All living organisms are made of **cells**. Some organisms, such as bacteria, are made of only one cell. Plants and animals are made of many cells. Scientists don't know how many cells are in the human body, but some estimate that the human body has trillions of cells.

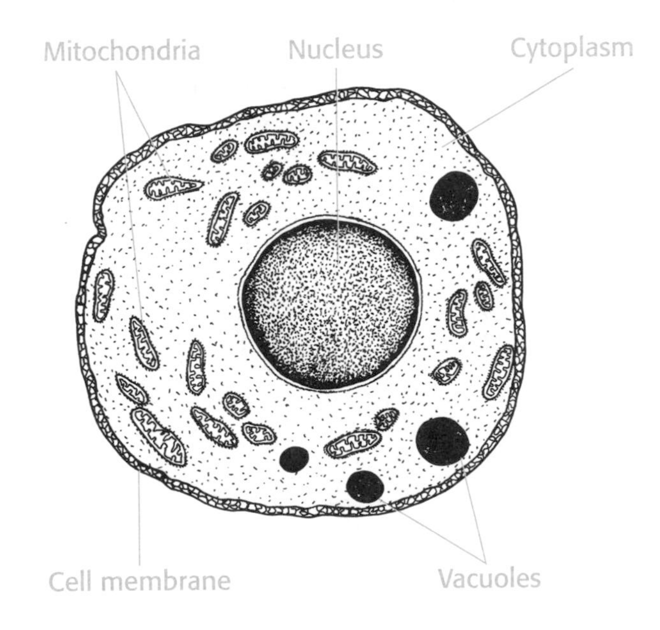

Animal cells come in many different shapes and sizes. But all animal cells have some things in common. The diagram to the left shows cell parts shared by all animal cells.

A **cell membrane** holds the contents of a cell together. Tiny holes in the membrane allow materials to pass through. The cell membrane controls the movement of materials into and out of the cell.

Inside the cell is a jellylike **cytoplasm**. The other cell parts float in the cytoplasm. The **nucleus** controls all the activities in the cell. Messages sent from the nucleus tell the other cell parts what to do. The **mitochondria** use food to create the energy the cell needs to carry out its activities. Most animal cells have several mitochondria. Food, water, or even waste waiting to be moved out of the cell are stored in **vacuoles**.

Show What You Know

Use the words below to complete each sentence.

mitochondria **vacuoles** **cell membrane** **nucleus**

1. The ____________________ is the control center of the cell.

2. The ____________________ controls what moves into and out of the cell.

3. The ____________________ provide(s) energy for the cell.

4. Water and nutrients are stored in ____________________.

LESSON 10

Plant Cells

How are plant and animal cells different?

Plants are made of cells, too. Like animal cells, plant cells come in many different shapes and sizes. But all plant cells have some things in common. The diagram below shows cell parts shared by all plant cells.

Like an animal cell, a plant cell has a **cell membrane.** The membrane holds the cell's contents together and controls the movement of materials into and out of the cell. Unlike an animal cell, a stiff **cell wall** surrounds the cell membrane. This firm cell wall allows plants to grow tall and still keep their shape.

Cell parts float in a jellylike **cytoplasm** that fills the cell. The cell's **nucleus** controls the cell's activities by sending chemical messages that tell the other cell parts what to do. As animal cells do, plant cells depend on **mitochondria** to create the energy the cell needs to do its work. This energy comes from food made in cell parts called **chloroplasts.** Chloroplasts contain **chlorophyll,** a green pigment that gives leaves and stems their color. Chlorophyll uses light energy from the sun to make food for the cell. Food, water, and even waste waiting to be moved out of the cell are stored in **vacuoles**. Plant cells usually have only a few vacuoles, or one large one.

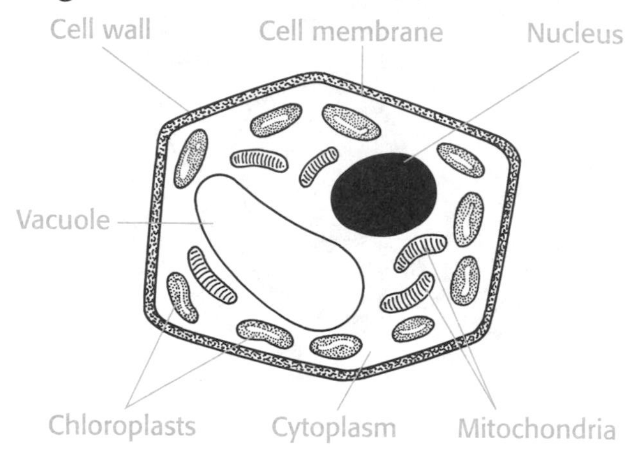

Show What You Know

Use the Venn diagram to name the cell parts present in animal cells, in plant cells, and in both kinds of cells.

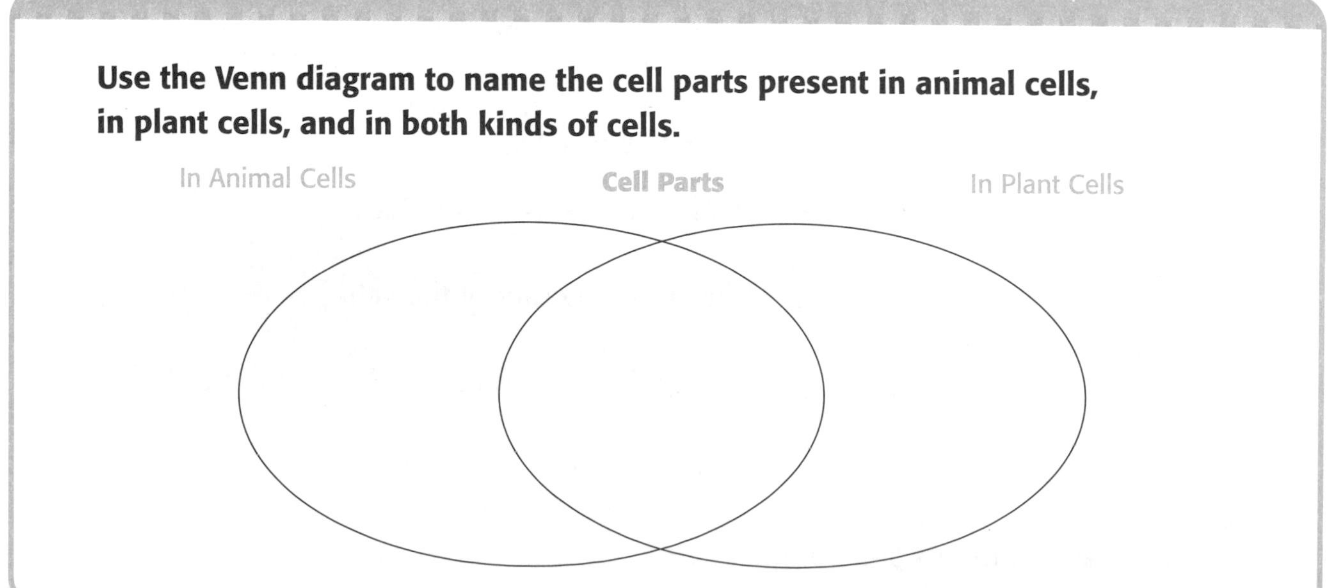

Tissues and Organs

What are tissues and organs?

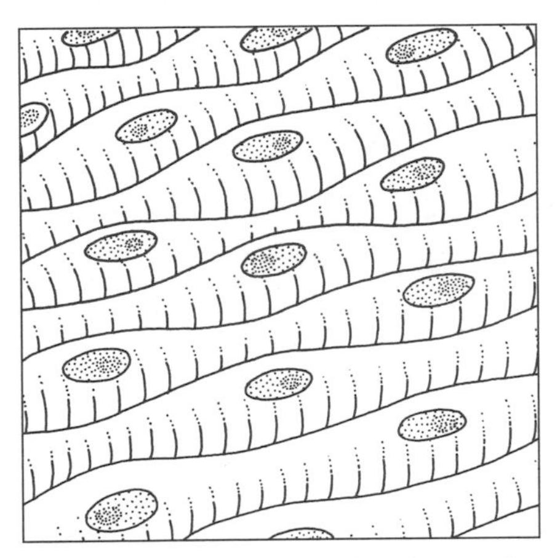

Muscle tissue is made of muscle cells, which are flat and layered. Muscle cells tighten and relax to move parts of a body.

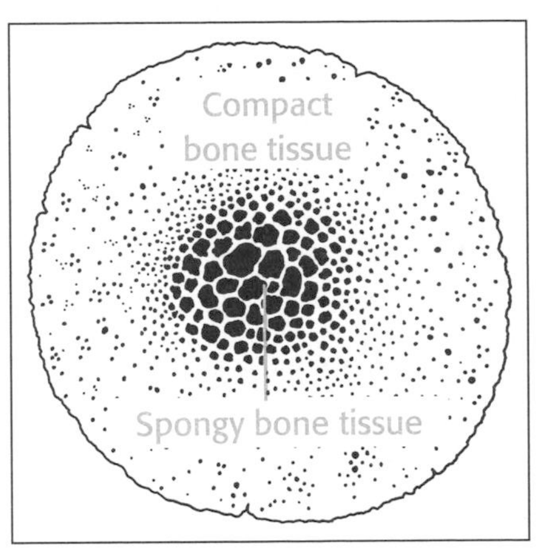

Bones are made of different kinds of **bone tissue**—spongy tissue inside the bone and hard tissue on the outside of the bone.

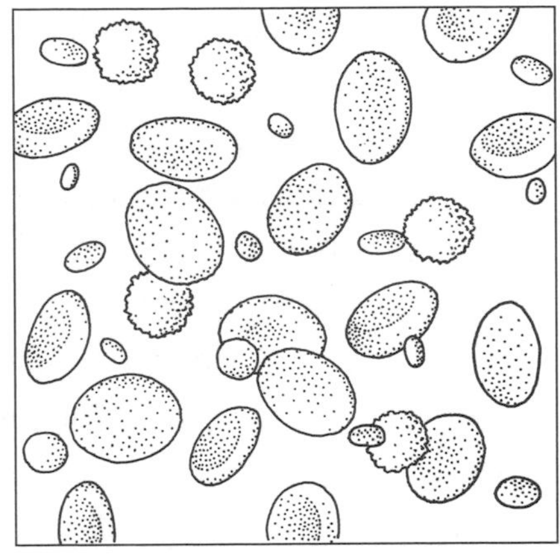

Blood tissue is made of different kinds of blood cells floating in a liquid. Each type of blood cell does a different job. Some carry oxygen. Others fight germs. There are also cell parts that help stop bleeding if a blood vessel is cut.

Although cells have things in common, they are not all alike. Plants and animals have different kinds of cells for different kinds of jobs.

A group of cells that work together to do a certain job form a **tissue**. Usually, the cells that make up a type of tissue are similar to each other. For example, muscle cells combine to form muscle tissue. Bone and blood are also examples of tissue. Each kind of tissue does a different job.

Tissues that work together to do a job form **organs.** The heart, stomach, brain, and lungs are all organs.

When organs work together to do a job, they form **organ systems.** The heart, blood, and blood vessels work together to move blood around the body. They form the circulatory system.

Show What You Know

Build a concept map explaining how tissues are organized in a body. Include the terms *cell, tissue, organ,* and *organ system* in your concept map.

LESSON 12

The Skeletal and Muscular Systems

How do the skeletal and muscular systems work together?

Your **skeletal system** supports your body and gives it shape. It also protects other body parts, such as your heart and lungs. Bones in your skeletal system meet at **joints.** Most joints let bones move in different directions.

Your skeletal system works with your **muscular system** to allow movement. **Skeletal muscles** pull on bones, letting you move your arms, legs, and other body parts when you want to.

Skeletal muscles aren't the only kind of muscle in the muscular system. **Heart muscles** work all by themselves to make your heart pump blood. You cannot control your heart muscles.

Another kind of muscle you cannot control is **smooth muscles.** These are found in organs like your stomach, lungs, and blood vessels. Smooth muscles move things like food and blood through organ systems.

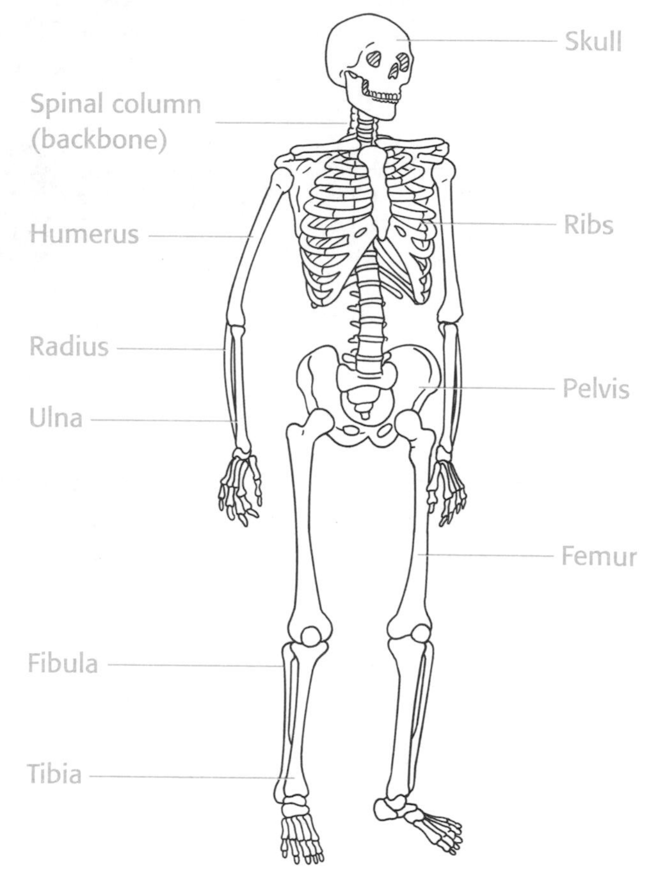

As an adult, your skeletal system will have 206 different bones.

Show What You Know

Complete the table to show the three types of muscles, if they can be controlled, and an example of where each can be found in the body.

	Muscle Type	What Is Its Job?	Can You Control It?
1.	____	____	____
2.	____	____	____
3.	____	____	____

The Digestive System

How do you get energy from food?

You must eat food to get the energy you need to live. Your **digestive system** is made of organs that work together to help you break down food into molecules that your cells can use to produce energy.

Your digestive system starts working when you put food in your **mouth**. As you chew, your teeth break the food into little pieces. The **saliva** in your mouth helps break the food down even more.

When you swallow, food travels through the **esophagus** to the stomach. In the **stomach,** the food is mixed together with more juices. Stomach muscles churn the mixture until the food is mostly liquid.

The liquid food then passes into the **small intestine**, where different juices are added to break the food down even more. Finally, food molecules move into tiny blood vessels in the small intestine. The blood carries them to the body's cells, where the cells use them to produce energy.

The parts of the food that can't be broken down any more travel to the **large intestine.** The large intestine takes the water out of the food. The leftover material is stored in the **rectum** before leaving the body as waste.

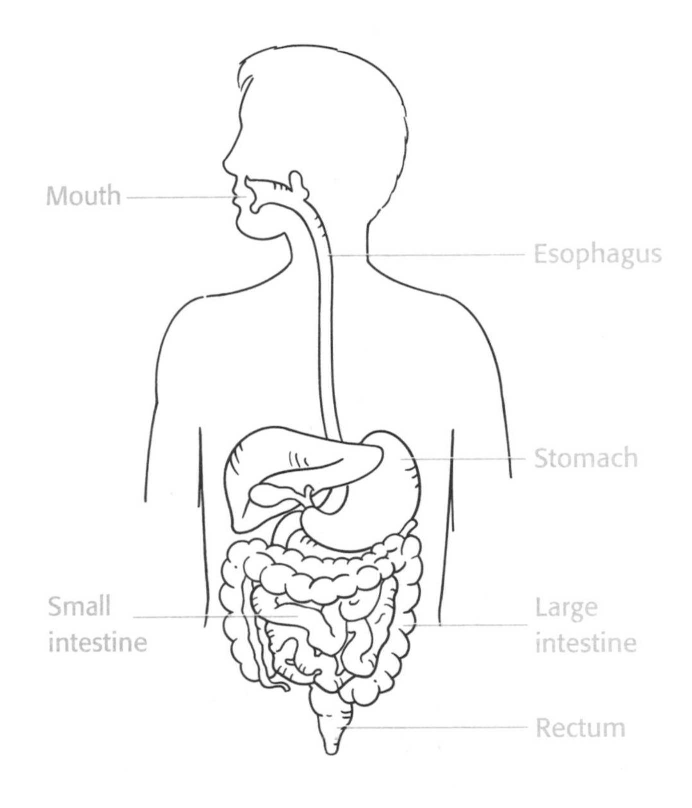

Show What You Know

Choose one organ in the digestive system. Describe how that organ helps digest food. Also explain how it prepares the food for the next organ in the system.

___

___

___

LESSON
14

The Nervous System

What controls your body systems?

The **nervous system** controls all of the body's systems The basic unit of the nervous system is a nerve cell, or **neuron.** Neurons receive signals from inside and outside the body. They also send signals to other neurons, to muscles, and to organs.

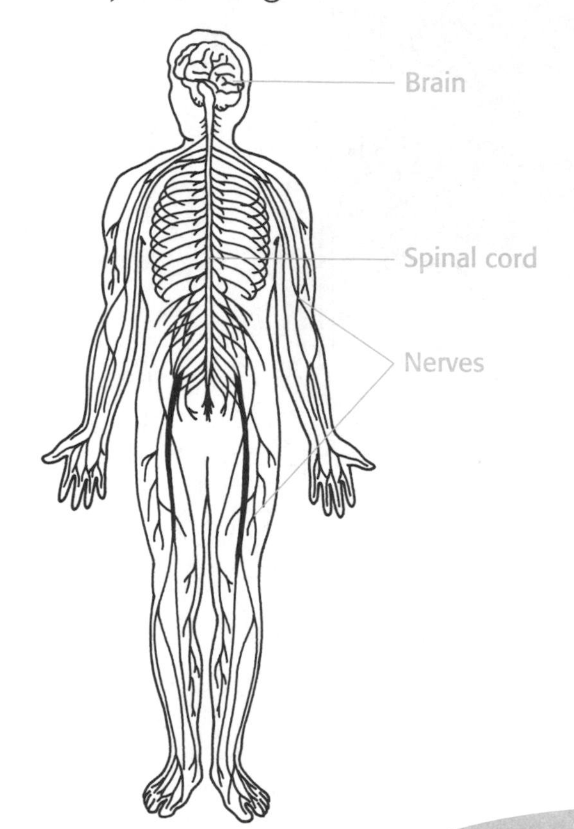

The nervous system has two parts. The **central nervous system** includes the spinal cord and brain. The other part of the nervous system includes nerves that run throughout your body.

The **spinal cord** runs from your brain down your spine. All the nerves in your body connect to the brain through the spinal cord. Messages sent between the body and the brain travel through the spinal cord.

The **brain** is the body's control center. It receives and processes information from the rest of the body. It then sends commands that tell the body how to respond.

Some of the information that goes to the brain comes from the **sense organs.** The sense organs include the eyes, ears, nose, tongue, and skin. They collect information about things you see, hear, smell, taste, or feel and send it to the brain.

Show What You Know

Imagine that you stepped on a pinecone while walking barefoot outside. Name the sense organ that would collect information about what you stepped on. Then explain where the organ would send the information, how it would get there, and what would happen after it arrived. Finally, draw the path the message would take on the diagram above.

____

____

The Respiratory and Circulatory Systems

How do the respiratory and circulatory systems work together?

To get energy from food, cells need oxygen. As they use oxygen, cells release carbon dioxide and water as waste. The **respiratory system** takes in oxygen for your cells to use. It also gets rid of waste. When you breathe in, you bring oxygen into your body. When you breathe out, you get rid of carbon dioxide and water vapor.

Air comes into the body through the **nose** and **mouth.** The air travels down a tube called the **trachea.** From there it goes into the **lungs** through the **bronchi.** The bronchi get smaller and smaller. Finally, the air reaches tiny air sacs. Here, oxygen from the air passes into tiny blood vessels. Carbon dioxide goes in the opposite direction. It leaves the blood and goes into the lungs to be breathed out.

Blood is part of the **circulatory system.** The job of the circulatory system is to move blood throughout the body. Blood brings oxygen and nutrients to the body's cells. It also carries away their waste. The **heart** is the "motor" that powers the circulatory system. Each beat of the heart pumps blood through the body.

Blood travels through **blood vessels.** **Arteries** are blood vessels that carry blood away from the heart. **Veins** return blood from the body to the heart. The very small blood vessels that connect arteries to veins are called **capillaries.**

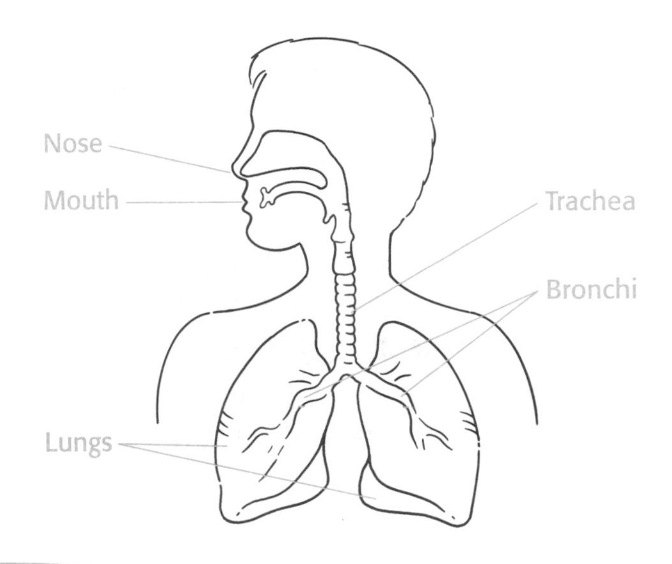

Show What You Know

How do the respiratory and circulatory systems work together to get oxygen to cells?

____

____

____

LESSON 16

Health

How can you keep your body healthy?

Making good food choices is an important part of staying healthy. Eating a variety of foods each day helps make sure your body gets the different nutrients it needs. Good food choices include snacks like fruits and vegetables instead of junk foods that are high in sugar and fat.

The Food Guide Pyramid breaks food into six different groups. It also suggests how many servings of each food group you should eat each day.

Exercise is also an important part of good health. Exercise does not have to be jumping jacks or aerobics. Doing things you enjoy, such as playing sports, can be good exercise. Even walking for several minutes each day is good exercise.

Good health also includes getting enough rest. Most people need at least eight hours of sleep each night. Children and teenagers need more. Not getting enough sleep can make it hard for you to concentrate. You may find yourself falling asleep during class or having trouble getting things done. Health experts have found that students who eat properly and get at least eight hours of sleep each night do better in school and sports than students who do not.

Food Guide Pyramid

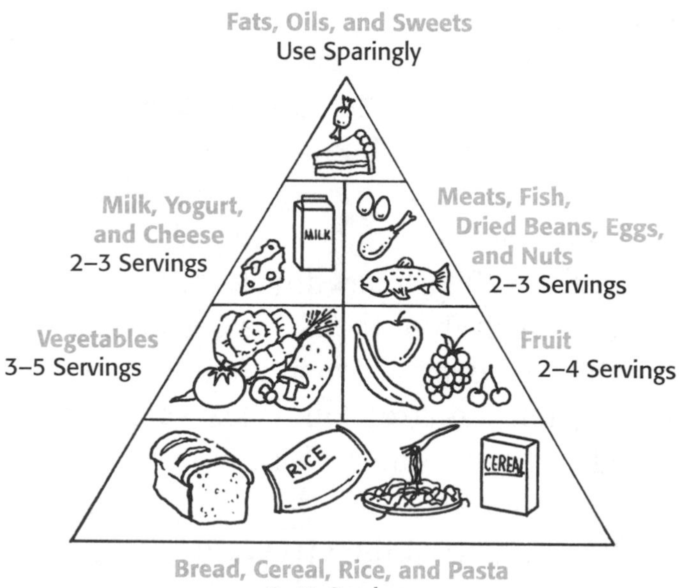

Show What You Know

1. According to the Food Guide Pyramid, how many daily servings of fruits and vegetables should be included as part of a healthy diet?

____

2. Circle all the food groups in the Food Guide Pyramid that were part of your breakfast this morning.

Fill in the letter to show your answer.

Use the diagram to answer Items 1-2.

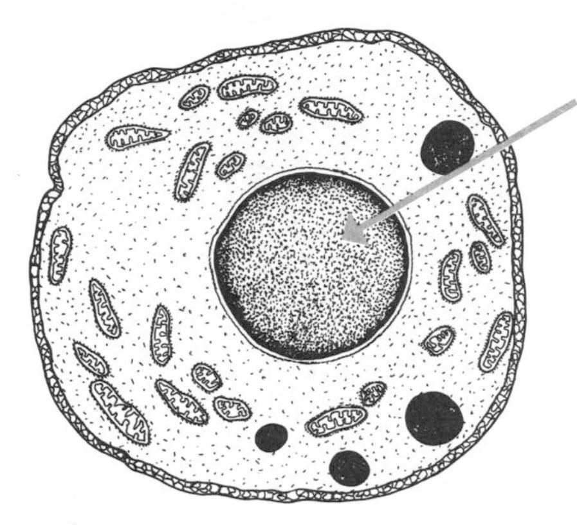

1. **Which cell part is the arrow pointing to?**

   Ⓐ nucleus

   Ⓑ vacuole

   Ⓒ mitochondria

   Ⓓ chloroplast

2. **The function of this cell part is to**

   Ⓐ store waste.

   Ⓑ provide energy for the cell.

   Ⓒ control the cell's activities.

   Ⓓ make food for the cell.

3. **Which cell part is *not* found in an animal cell?**

   Ⓐ mitochondria

   Ⓑ chloroplast

   Ⓒ cell membrane

   Ⓓ vacuole

• • LIFE SCIENCE TEST B • •

4. Tissues are made of groups of similar

Ⓐ cells.

Ⓑ muscles.

Ⓒ organs.

Ⓓ systems.

5. Which letter in the diagram indicates the part of the digestive system where molecules are absorbed into the blood?

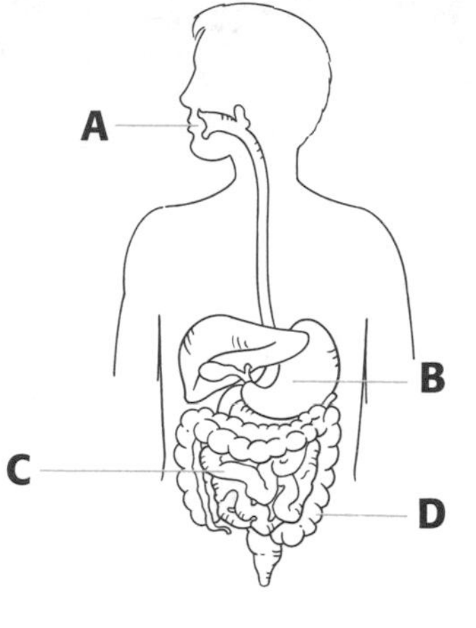

Ⓐ A

Ⓑ B

Ⓒ C

Ⓓ D

6. Which organ is the control center of the body?

Ⓐ the heart

Ⓑ the brain

Ⓒ the eyes

Ⓓ the stomach

Short Response

7. How does the respiratory system help the body get energy from food?

____

____

____

____

Ecosystems

What makes up an ecosystem?

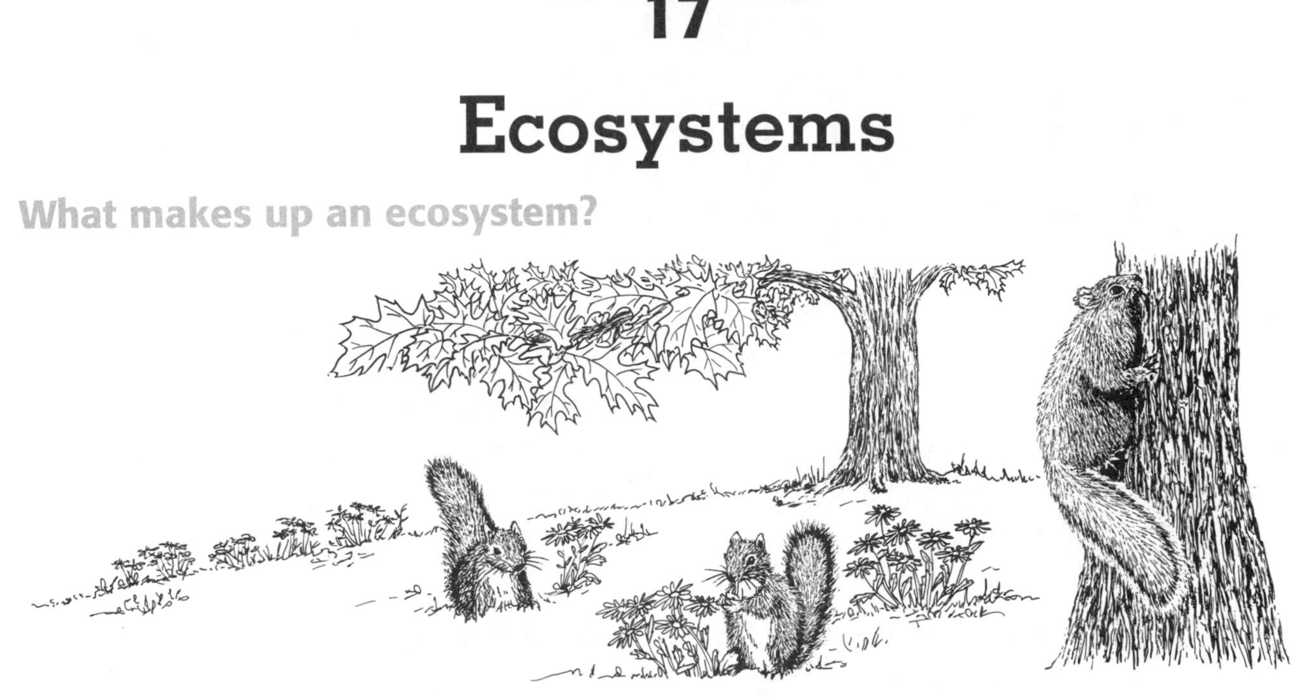

A **species** is a single kind of organism that can mate with other organisms like itself and have offspring. A gray squirrel is one kind of species. A red oak is another. Organisms of the same species that live in an area make up a **population.** Together, the populations living in an area make up a **community** of organisms. A community along with the nonliving parts of its environment form an **ecosystem.**

The specific environment that meets all the needs of an organism is that animal's **habitat.** The habitat of a squirrel can be a field, a city park, or even a backyard. Within its habitat, the squirrel does a number of things. It eats acorns, builds a nest in the trees, and acts as food for foxes that visit the field. All the things an organism does make up its role, or **niche,** in the habitat.

Show What You Know

Make a concept map to explain the parts of an ecosystem. Include these terms in your map: species population community

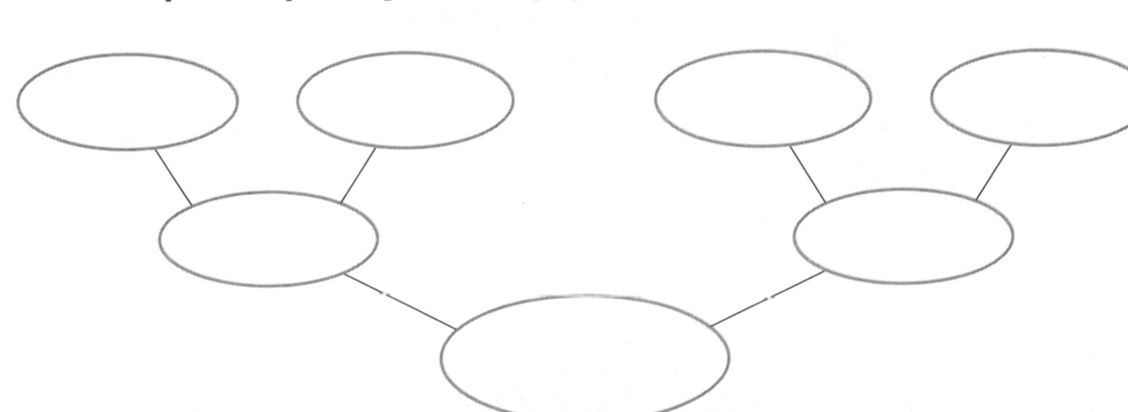

LESSON 18

Nonliving Parts of Ecosystems

What are the nonliving parts of an ecosystem?

Sunlight, water, soil, and air are all nonliving parts of an ecosystem. They determine what kinds of organisms can live in an area.

Living things in an ecosystem can be affected when nonliving parts of an ecosystem change. For example, if an ecosystem receives less than normal rainfall one year, a pond in the ecosystem could dry up. Water plants that grow in the pond could die. Birds that eat the water plants would leave the ecosystem in search of other water plants to eat.

Nonliving Parts of an Oak Forest Ecosystem

Sunlight Energy for all living things comes from sunlight. Plants use sunlight to make their own food through photosynthesis. Many animals eat plants to live.

Water All living things need water to live. Plants also need water to make food.

Soil Most plants grow in soil. The soil provides the plants with the water and nutrients they need. Animals burrow into soil for shelter.

Air Air is made of gases organisms need to live, including carbon dioxide and oxygen. Plants use carbon dioxide to make food. Both plants and animals use oxygen to get energy from food. These two gases are always cycling between organisms and the environment.

Show What You Know

Which ecosystem do you think is home to more organisms—one with dry, sandy soil or one with moist, nutrient-rich soil? Explain your answer.

____

____

____

Lesson 19

Roles in an Ecosystem

What are the roles of living things in an ecosystem?

Life Science

Plants are **producers.** They use water, carbon dioxide, and the energy in sunlight to make their own food. Animals are **consumers.** They must eat plants or other animals as food. **Decomposers** like bacteria and fungi get energy by feeding on the bodies of dead plants and animals. As they break down dead organisms, the decomposers release valuable nutrients back into the soil.

Three kinds of consumers are herbivores, carnivores, and omnivores. Animals that eat plants for food are **herbivores.** Animals that eat other animals for food are **carnivores.** Animals that eat plants and other animals are **omnivores.**

An animal that hunts and catches another animal is a **predator.** The animal that the predator catches is its **prey.** A **parasite** is an organism that lives off another organism. The organism that the parasite lives off is the **host.**

A Pond Ecosystem

Show What You Know

Name one producer and one consumer in the picture above.

Producer: ____________________

Consumer: ____________________

LESSON 20

Interactions in an Ecosystem

How do living things in an ecosystem depend on each another?

Animals depend on plants for food and shelter. Deer eat grasses and shrubs. Birds make their homes inside hollow trees. Lions nap in the cool shade of a tree. Without plants, animals could not live.

Plants need animals, too. Many plants that reproduce with flowers depend on animals for pollination. **Pollination** is the movement of **pollen** from the **stamens** of a flower to the **pistil** of the same flower, or another flower. A flower needs to be pollinated before it will form seeds. Seeds are what the plant uses to reproduce. Animals that help move pollen include bees, butterflies, flies, birds, and bats.

Some plants also depend on animals to spread their seeds. When an animal eats the fruit of a plant, the seeds in the fruit are often not digested. Instead, the seeds leave the animal's body with its wastes. They can then grow where they are dropped. Sometimes seeds stick to an animal's fur as it walks by a plant. The seeds then drop off and grow somewhere else.

As a bee explores a flower, it moves pollen from the stamens to the pistil.

Show What You Know

Draw a set of pictures to show how an animal can spread the seeds of a plant just by walking by the plant.

Food Chains

How does energy move through an ecosystem?

The movement of energy from one organism to another in an ecosystem can be shown in a **food chain.** A food chain starts with a **producer.** Plants use sunlight to make food. They use some of the food they make to live. The rest is stored in the leaves, stems, and roots of the plant.

The next organism in the food chain is a **consumer** that eats the plant. Then comes another consumer that eats the first consumer, and so on. These consumers use some of the energy they get by eating to live. The rest is stored in their bodies. Finally, **decomposers** get energy by feeding on the bodies of dead plants and animals.

Only about 10% of the energy in each organism's body is passed on to the organism that eats it. So, at each step in the food chain, there is less food energy available. That explains why there are fewer sharks than small fish in an ocean ecosystem.

An Ocean Ecosystem Food Chain

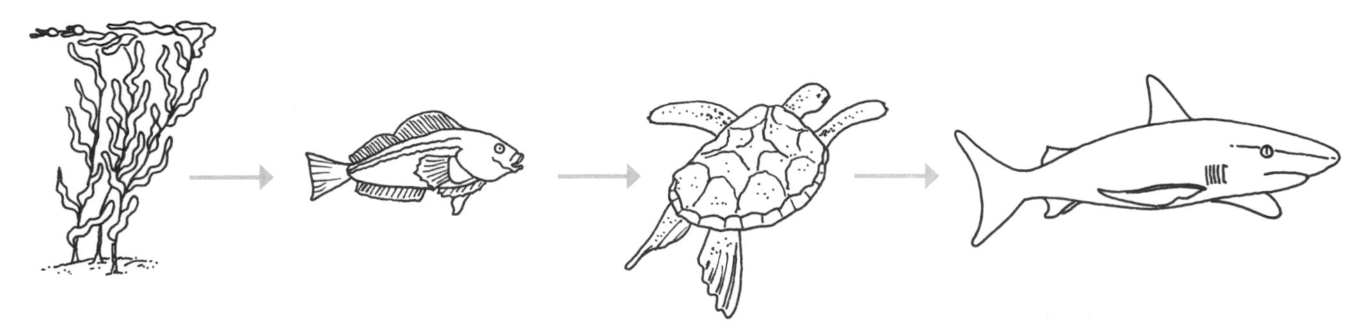

Show What You Know

1. The following words name things in a desert ecosystem. Arrange the words to form a food chain. Then, add arrows to show how energy moves through the food chain. **mouse** **grass** **sun** **bacteria** **coyote**

2. Why are there fewer coyotes than mice in a desert ecosystem?

___

___

LESSON 22

Food Webs

How do food chains form a food web?

A food chain shows how energy moves between one set of living things in an ecosystem. But, the movement of energy within an ecosystem is not really this simple. That's because organisms often form part of several different food chains. For example, a duck might be food for a snake. But, it could also be food for an eagle. All the food chains in an ecosystem together make up a **food web.**

A change in one part of a food web can affect all the organisms in the ecosystem. Let's say some of the producers in an ecosystem died out. Herbivores such as prairie dogs would have less to eat. Some would leave to find food. Others would die. Then, the consumers that eat prairie dogs wouldn't have enough food. They would leave or die, too. In time, all the organisms in the ecosystem could be affected.

A Prairie Ecosystem Food Web

Show What You Know

What could happen in the prairie ecosystem if the prairie dog population suddenly decreased a lot?

____

____

____

Fill in the letter to show your answer.

1. **An organism's special role in an environment is its**

   Ⓐ population.

   Ⓑ habitat.

   Ⓒ niche.

   Ⓓ food web.

2. **Which nonliving part of an ecosystem do plants need in order to make food?**

   Ⓐ water

   Ⓑ carbon dioxide

   Ⓒ sunlight

   Ⓓ all of the above

3. **All the members of a species living together in an area form a**

   Ⓐ population.

   Ⓑ community.

   Ⓒ ecosystem.

   Ⓓ food web.

4. **Which of the following is at the end of every food chain?**

   Ⓐ producer

   Ⓑ consumer

   Ⓒ decomposer

   Ⓓ herbivore

• • LIFE SCIENCE TEST C • •

Use the diagram to answers Items 5-7.

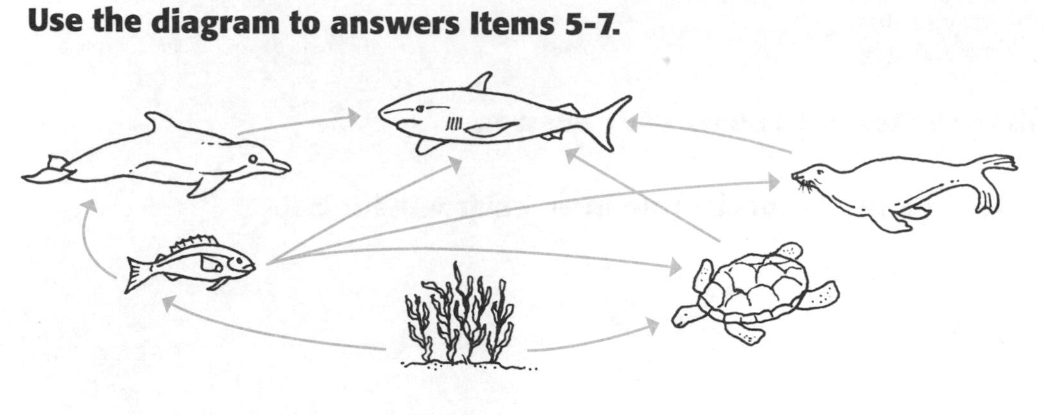

5. **What is shown in the diagram?**

   Ⓐ a food chain

   Ⓑ a food web

   Ⓒ a population

   Ⓓ a niche

6. **What is the sea turtle's role in the diagram?**

   Ⓐ producer

   Ⓑ herbivore

   Ⓒ consumer

   Ⓓ decomposer

Short Response

7. **What would happen if all the dolphins in this ecosystem died?**

____

____

____

Classifying Organisms

How do we classify organisms?

There are millions of different kinds of organisms in the world. Scientists **classify,** or group, organisms based on their **traits,** or characteristics. Traits include the number and kinds of cells the organisms are made of. They also include how organisms get food, whether they move, and how they reproduce.

A **kingdom** is the largest classification group. There are five kingdoms—animals, plants, fungi, protists, and bacteria. All organisms belong to one of these kingdoms. The smallest grouping in a kingdom is a **species.** It includes only one kind of organism. A Venus's flytrap, for example, is a species within the plant kingdom.

Animals Animals are made of many cells. Animal cells do not have **cell walls** or **chloroplasts.** Animals cannot make their own food. They must eat other organisms for food. Animals can move around. Some animals reproduce by laying eggs. Others give birth to live young.

Plants Plants are also made of many cells. Plant cells have **cell walls** and **chloroplasts.** Plants use sunlight to make their own food. They cannot move around on their own. Plants reproduce using **seeds** or **spores.**

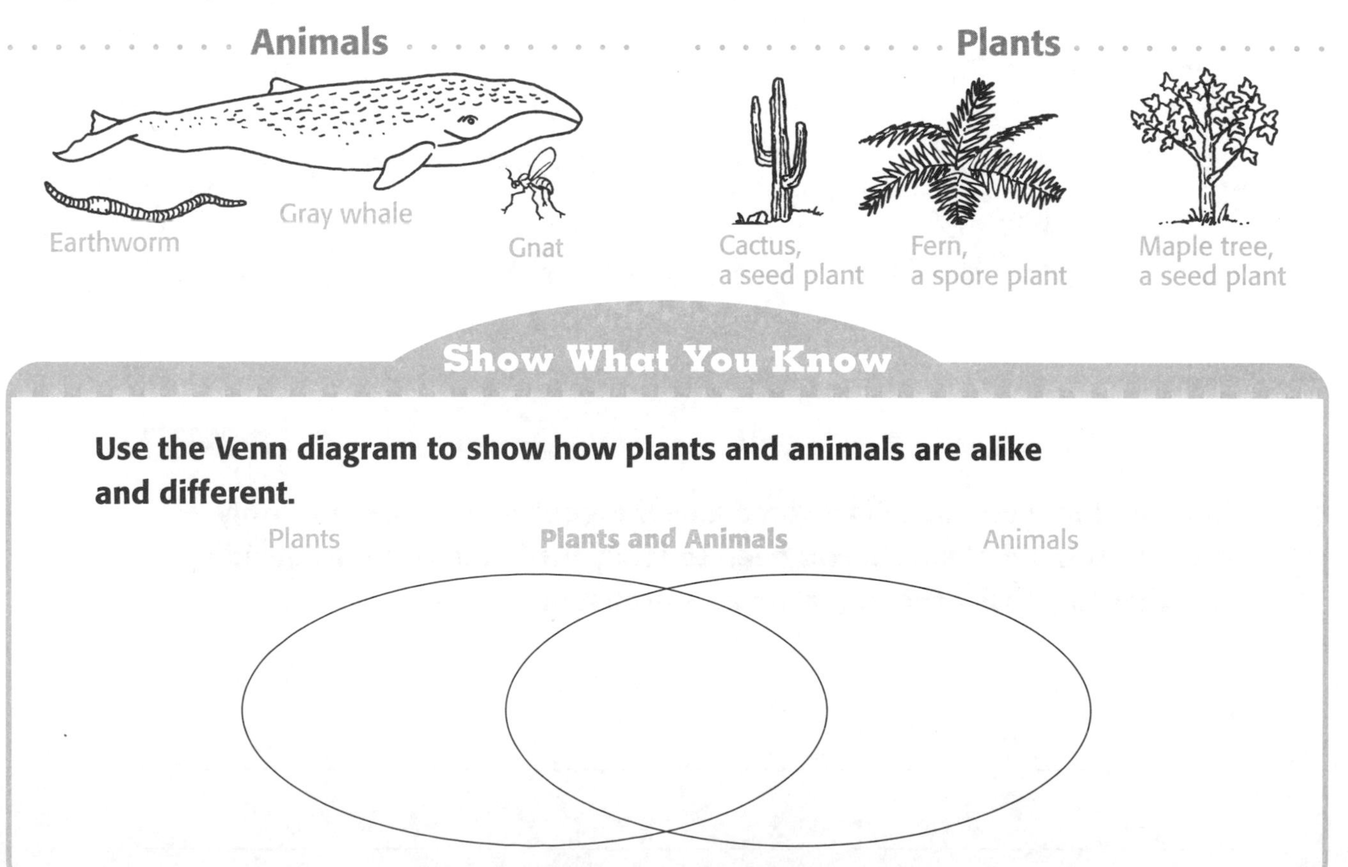

Use the Venn diagram to show how plants and animals are alike and different.

LESSON 24

Kingdoms

What are the other kingdoms?

Fungi Some fungi are made of one cell. Others are made of many cells. All fungi cells have a **nucleus** and a **cell wall.** Fungi cannot move around or make their own food. They are **decomposers,** and so they feed on dead plants and animals. Fungi reproduce using **spores.** A mushroom is a fungus.

Protists Most protists are made of only one cell. But some are made of many simple cells strung together. All protist cells have a nucleus. The amoeba is a one-celled protist. It moves around and feeds on other organisms for food. Algae are protists that have many cells. They cannot move around. They use sunlight to make their own food.

Bacteria Bacteria are made of only one cell. That cell does not have a nucleus. Some bacteria can use sunlight to make their own food. Others must get the energy they need from other organisms. Scientists classify bacteria by their shape—round, spiral, or rod. Some bacteria cause diseases, like strep throat. But most bacteria are useful, like the bacteria used to make yogurt.

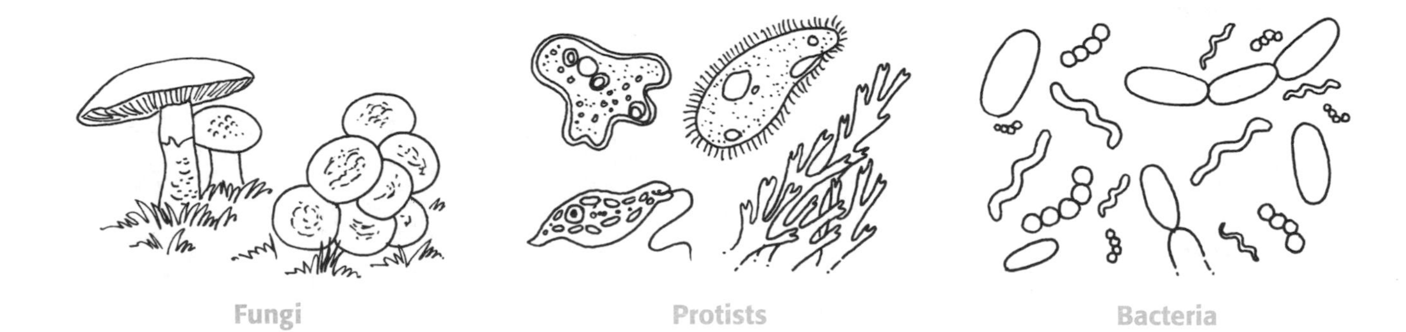

Fungi Protists Bacteria

Show What You Know

Imagine that you have discovered a new organism. It is made of only a single cell. What would you need to study further in order to decide whether to classify the organism as a protist or a bacterium?

____

____

____

Invertebrates

What makes an animal an invertebrate?

An **invertebrate** is an animal without a backbone. Invertebrates are classified into many groups. They include sponges, jellyfish, worms, mollusks, and arthropods.

Sponges eat by pulling ocean water through the holes in their bodies and filtering out the bits of food in the water. Water moving out the top of the sponge carries away wastes. **Jellyfish** have only one body opening. Food they capture with their tentacles enters the body opening. Later, wastes are passed out of the same opening. **Earthworms** have two body openings, one for eating and one for getting rid of wastes.

Mollusks have soft bodies, but most are covered by a shell that protects them. Mollusks have a strong foot that they use for holding on to rocks, burrowing in sand, or moving across surfaces.

Arthropods include crabs, insects, and spiders. Arthropods have an **exoskeleton,** which is a hard outer covering that protects and supports the body. They also have legs with joints.

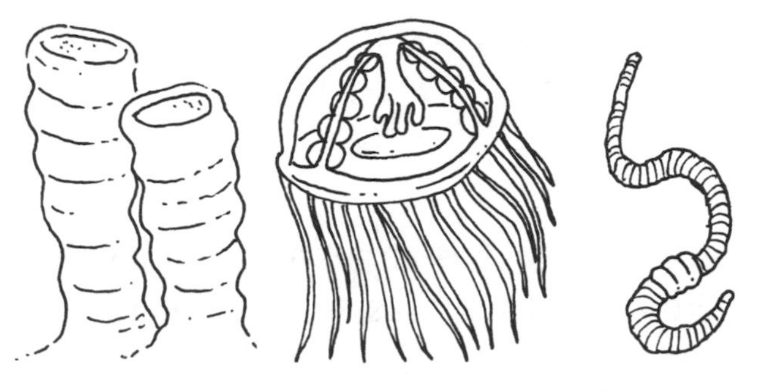

Sponge Jellyfish Earthworm

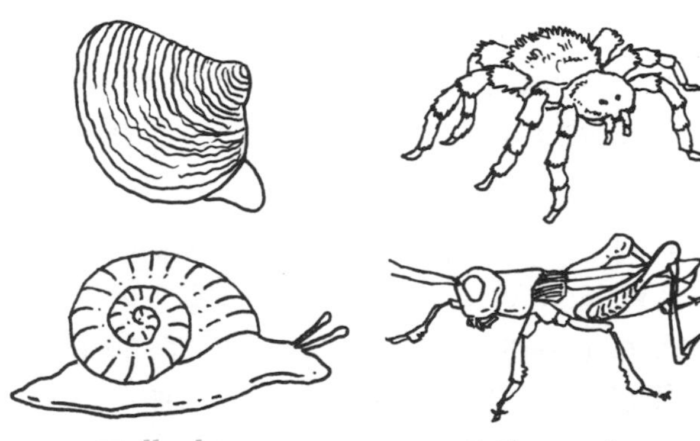

Mollusks Arthropods

Show What You Know

Sponges don't move and they don't look like animals. What makes them animals? Explain why they aren't classified as plants.

____

____

____

LESSON 26

Vertebrates

What makes an animal a vertebrate?

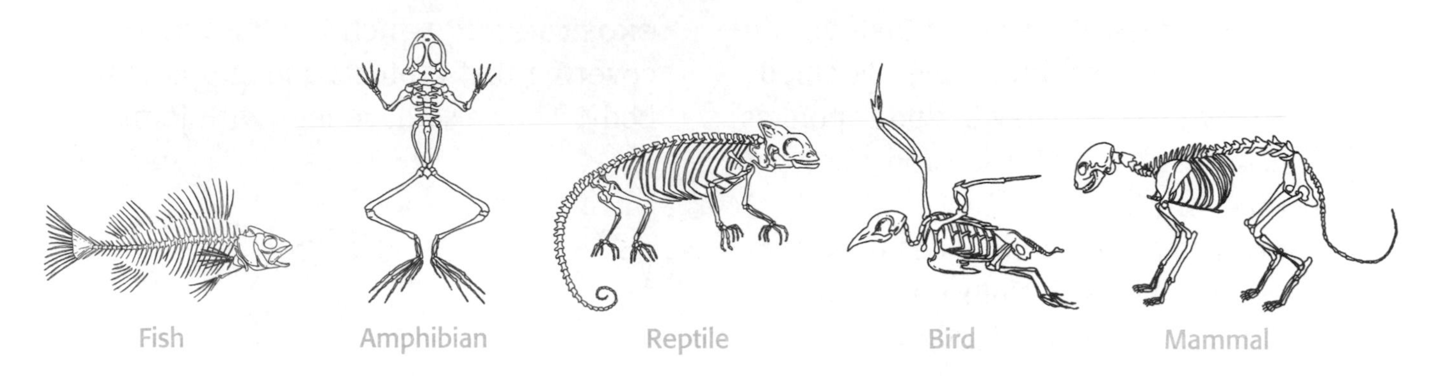

A **vertebrate** is an animal that has a backbone. Vertebrates are classified into five main groups—fish, amphibians, reptiles, birds, and mammals.

Fish live in water. Most have scales and fins. All have **gills** for getting oxygen out of the water. Most fish lay eggs that don't have a shell. Others, like sharks, have live young.

Amphibians are born in water. They have gills when they are young, but develop lungs and live on land as adults. Their skin is thin and moist. Most amphibians lay jellylike eggs in water. Frogs and toads are amphibians.

Reptiles have scales that protect the body. They lay eggs with a leathery covering that protects the eggs and keeps them from drying out. Reptiles breathe with lungs. Snakes, turtles, and lizards are reptiles.

Birds have feathers that keep them warm. Most birds use their feathers to help them fly. Birds lay eggs that contain food for the developing bird. The shell protects the egg from drying out. Robins and penguins are birds.

Mammals produce milk that they use to feed their young. Mammals have hair that keeps them warm. Most mammals give birth to live young. Elephants and bats are mammals. You are a mammal.

Show What You Know

Match each body covering with the group of animals it can be found on.

1. feathers
2. hair
3. scales
4. thin, moist skin

a. fish and reptiles
b. mammals
c. amphibians
d. birds

A Multiple Choice

Fill in the letter to show your answer.

1. **What do plants and animals have in common?**

   Ⓐ Both can make their own food.

   Ⓑ Both have to eat other organisms for food.

   Ⓒ Both are made of cells.

   Ⓓ Both can move around.

2. **A scientist found organisms that each had one cell and no nucleus. What are they?**

   Ⓐ animal

   Ⓑ plant

   Ⓒ protist

   Ⓓ bacteria

3. **Which of the following organisms reproduce using spores?**

   Ⓐ plants and animals

   Ⓑ plants and fungi

   Ⓒ plants and protists

   Ⓓ plants and bacteria

4. **Which of the following traits is *not* used to classify organisms?**

   Ⓐ whether they move

   Ⓑ where they live

   Ⓒ how they get food

   Ⓓ how they reproduce

• • LIFE SCIENCE TEST D • •

Use the diagram to answer Item 5 below.

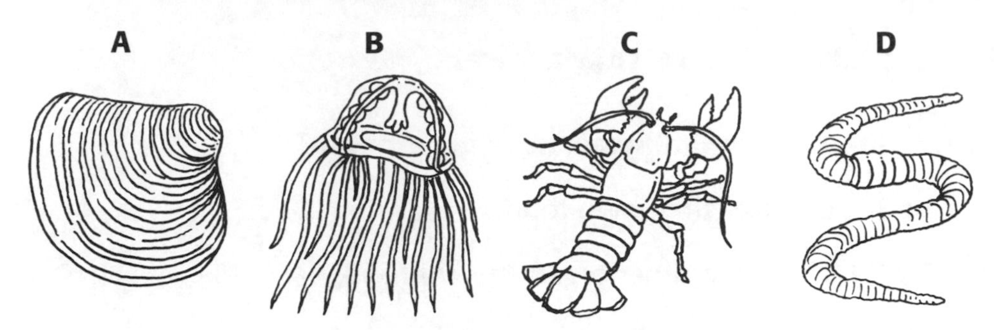

5. **Which of the organisms above has an exoskeleton?**

   (A) clam

   (B) jellyfish

   (C) lobster

   (D) earthworm

6. **One way that a fish, a snake, and a bird are alike is that they all have**

   (A) legs.

   (B) hair.

   (C) lungs.

   (D) backbones.

Short Response

7. **What kind of eggs would you expect an animal that lives in the desert to lay? Explain your answer.**

____

____

____

____

Check your answers to questions on Life Science Test A on pages 13-14.

Multiple Choice

1. **Photosynthesis creates a waste product that is needed by almost all living organisms. This waste product is**

   B oxygen.

   Almost all organisms need oxygen to release the energy in food.

2. **Black bears can spend up to 100 days during the winter without eating or drinking. This kind of adaptation is called**

   C hibernation.

   Hibernation is an inherited behavioral adaptation. Animals hibernate to stay alive during the winter, when food is scarce and temperatures are very low.

3. **Which bird beak would be best adapted for scooping small fish from the water?**

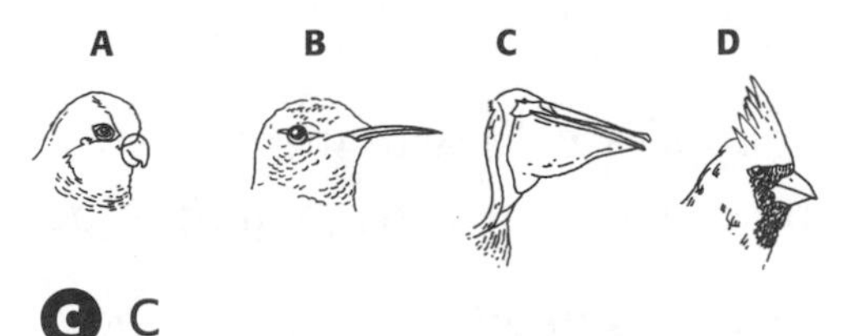

   C C

   The pelican uses the pouch on its lower bill to skim fish from the water.

4. **The plant structures in which photosynthesis usually takes place are the**

   A leaves.

   Leaves are the parts of the plant in which most photosynthesis takes place.

5. **Flowering plants reproduce by forming**

   D seeds.

   Seeds form in the female part of the flower and are surrounded by a fruit.

Short Response

6. **Explain how light and gravity affect the growth of plants.**

   Plant stems grow toward light and away from the pull of gravity. Plant roots grow toward the pull of gravity. These behaviors, or adaptations, explain why stems grow upward and roots grow downward.

   Plant stems grow upward in response to light and gravity. Roots grow downward in response to gravity.

Check your answers to questions on Life Science Test B on pages 23-24.

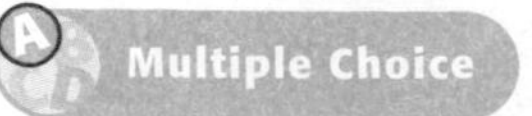

1. **Which cell part is the arrow pointing to?**

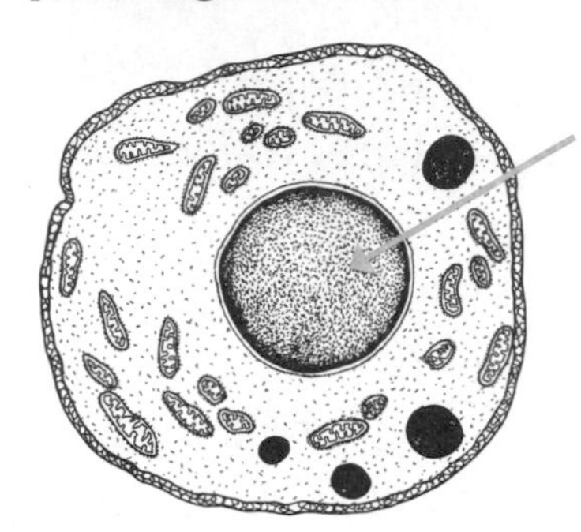

Ⓐ nucleus

The nucleus is large, round, and found near the center of the cell.

2. **The function of this cell part is to**

Ⓒ control the cell's activities.

The nucleus is the control center of the cell. It directs all of the activities of the cell.

3. **Which cell part is *not* found in an animal cell?**

Ⓑ chloroplast

Plant cells use chloroplasts to make food. Since animals do not make their own food, their cells do not have chloroplasts.

4. **Tissues are made of groups of similar**

Ⓐ cells.

Groups of similar cells that work together make up tissues.

5. **Which letter in the diagram indicates the part of the digestive system where molecules are absorbed into the blood?**

Ⓒ C

Molecules are absorbed into the blood in the small intestine. In the diagram, the letter C identifies the small intestine.

6. **Which organ is the control center of the body?**

Ⓑ the brain

The brain controls the functions of all the other body systems.

Short Response

7. **How does the respiratory system help the body get energy from food?**

It supplies cells with the oxygen they need to get energy from food.

Oxygen is needed to release the energy from food. The respiratory system delivers oxygen to cells.

Check your answers for Life Science Test C on pages 31-32.

Multiple Choice

1. **An organism's special role in an environment is its**

   (C) niche.

   An animal's niche is the special role it plays in its habitat.

2. **Which nonliving part of an ecosystem do plants need in order to make food?**

   (D) all of the above

   Plants need water, carbon dioxide, and sunlight for photosynthesis, the process plants use to make food.

3. **All the members of a species living together in an area form a**

   (A) population.

   A population is made up of all the members of one species living in one area.

4. **Which of the following is at the end of every food chain?**

   (C) decomposer

   Decomposers break down the remains of dead plants and animals. This returns nutrients to the soil.

5. **What is shown in the diagram?**

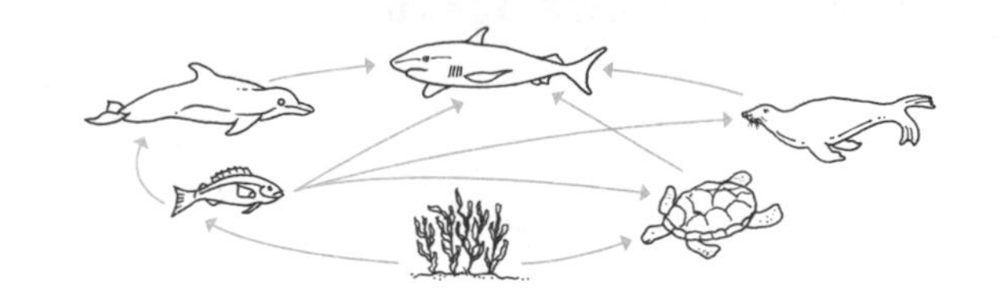

   (B) a food web

   A food web shows how different food chains overlap in an ecosystem.

6. **What is the sea turtle's role in the diagram?**

   (C) consumer

   An animal that eats a plant or another animal for food is a consumer.

Short Response

7. **What would happen if all the dolphins in this ecosystem died?**

   The fish population might increase because they have fewer predators. The shark population might decrease because they have fewer prey.

   A change in one part of a food web can affect all of the other organisms in the web.

Check your answers for Life Science Test D on pages 37-38.

Multiple Choice

1. **What do plants and animals have in common?**

   Ⓒ Both are made of cells.

   Both plants and animals are made of many cells.

2. **A scientist found organisms that each had one cell and no nucleus. What are they?**

   Ⓓ bacteria

   Bacteria are made of one cell, and that cell has no nucleus.

3. **Which of the following organisms reproduce using spores?**

   Ⓑ plants and fungi

   Plants reproduce using seeds or spores. Fungi reproduce using spores.

4. **Which of the following traits is *not* used to classify organisms?**

   Ⓑ where they live

   Organisms are not classified according to where they live.

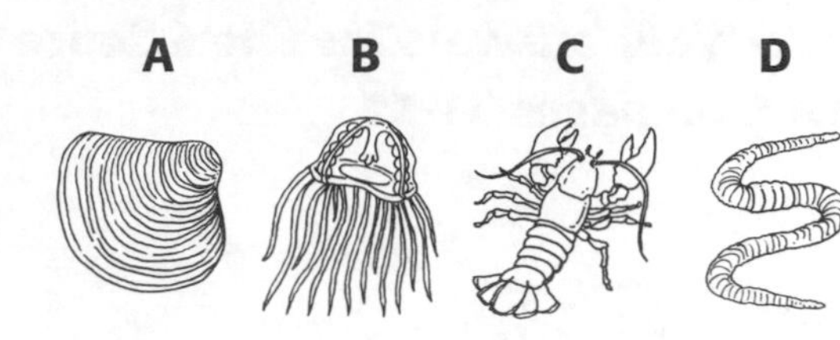

5. **Which of the organisms above has an exoskeleton?**

   Ⓒ lobster

   Arthropods have an exoskeleton. A lobster is an arthropod.

6. **One way that a fish, a snake, and a bird are alike is that they all have**

   Ⓓ backbones.

   Fish, reptiles, and birds are all vertebrates—animals with backbones.

7. **What kind of eggs would you expect an animal that lives in the desert to lay? Explain your answer.**

   A desert animal would lay eggs with leathery coverings so the eggs wouldn't dry out.

   A leathery covering would protect the eggs from drying out in the dry desert.

Rocks

How are rocks classified?

Rocks are mixtures of minerals and other earth materials. The three main types of rocks are igneous, sedimentary, and metamorphic. They are classified two ways—by their chemical makeup and by how they form.

Igneous rocks form when melted rock cools and hardens. Some igneous rocks form from **magma** beneath Earth's surface. They cool slowly, forming large mineral **crystals**. Other igneous rocks form from **lava** that flows onto Earth's surface. They cool quickly, forming small mineral crystals.

Sedimentary rocks form from bits of **sediment** that are packed together and cemented. The sediments are often laid down in layers. Some sedimentary rocks are made of shells and **fossils.** Scientists study fossils to learn about Earth's history.

Heat and pressure from forces inside Earth form **metamorphic rocks.** Sometimes, the heat and pressure arrange the minerals in the rock into layers. Other times, heat and pressure cause the rock's mineral grains to grow larger.

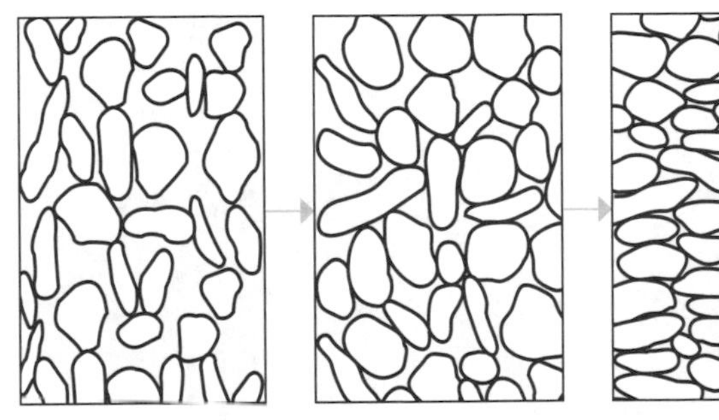

Sediments are tightly packed together to form sedimentary rock.

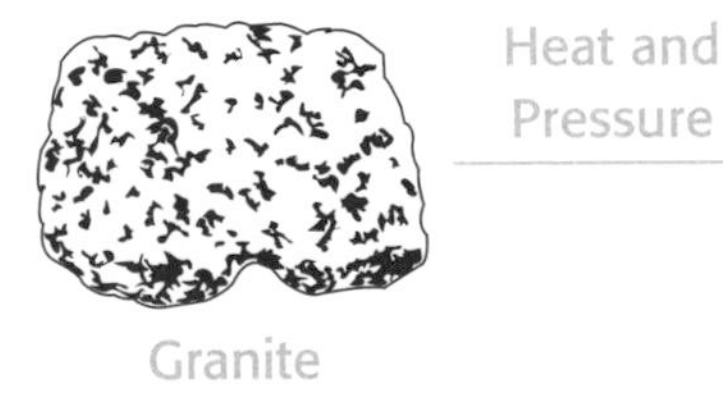

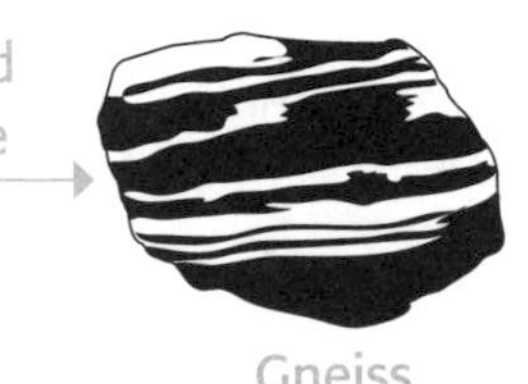

Heat and pressure change granite, an igneous rock, into gneiss, a metamorphic rock.

Show What You Know

Rocks A and B are igneous rocks. Rock A is made of large crystals. Rock B is made of very small crystals. Which rock formed beneath Earth's surface? Explain how you know.

____

____

____

LESSON 28

The Rock Cycle

What processes make up the rock cycle?

Over time, igneous, sedimentary, and metamorphic rocks can change from one rock type to another. The **rock cycle** shows the processes that both form and change rocks.

The arrows in the diagram show the different paths that a rock can take through the rock cycle. **Weathering** and **erosion,** for example, break down rock and carry away sediments. In time, the sediments can cement to form a sedimentary rock. These processes take place on Earth's surface.

Forces beneath Earth's surface change rock, too. For example, factors such as pressure and heat can change a sedimentary rock into a metamorphic rock. If that metamorphic rock gets hot enough to melt, it could later cool and harden into an igneous rock.

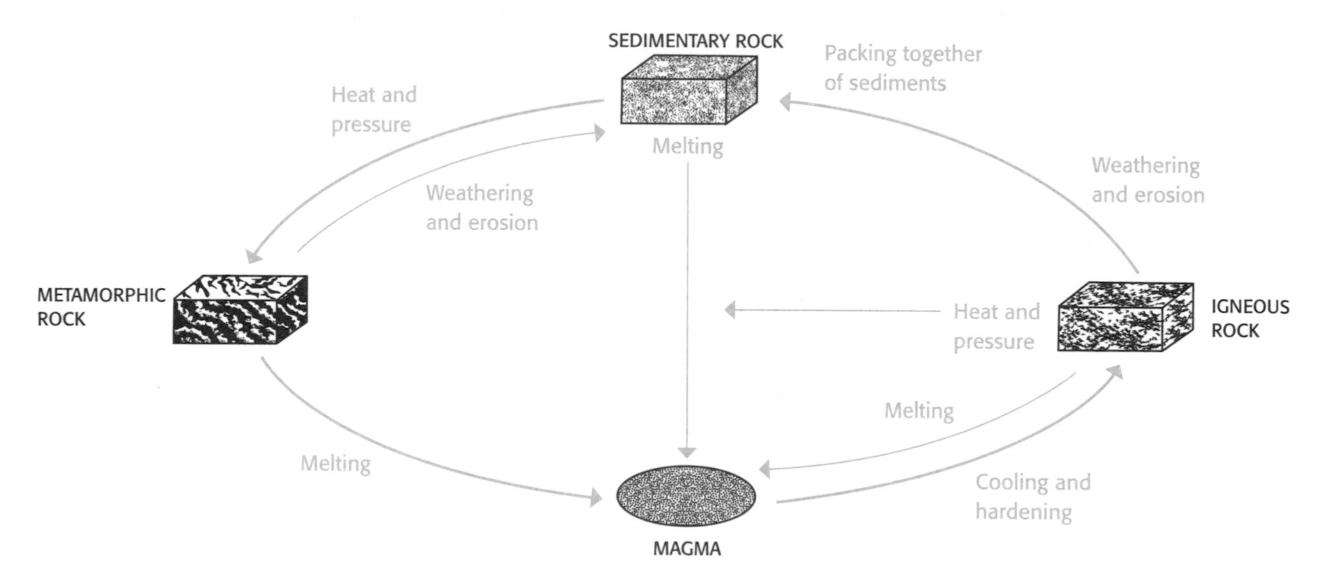

There is no beginning or end to the rock cycle. Rocks are constantly changing into other rocks.

Show What You Know

Draw lines to connect each rock type to the processes that created it.

1. igneous rock	a. weathering and erosion
2. metamorphic rock	b. hardening and cooling
3. sedimentary rock	c. heat and pressure
4. sediments	d. packing together and cementing of sediments

Minerals

How can minerals be identified?

All **minerals** share certain properties. They are solids with unique crystal structures. They are formed by natural processes. They are not made of the remains of once-living things. Instead, they are made of **elements** and **compounds** that join together in different ways.

In addition to these shared properties, each mineral has properties that make it different. Calcite, for example, is usually light in color. Color is one way to identify a mineral. **Luster** describes the way a mineral reflects light. Some minerals are shiny, while others look dull.

Hardness refers to how easy or hard it is to scratch a mineral. On a scale of 1 to 10, talc is the softest mineral and diamond is the hardest.

Minerals can also be identified by their streak. **Streak** is the color left by a mineral when it is rubbed across a tile. Pyrite and gold, for example, are both yellow in color. But pyrite's streak is greenish black, and gold's streak is yellow.

Mineral Properties

Mineral	Luster	Hardness	Color	Streak
Calcite	glassy	3	colorless, white, yellow	white
Feldspar	glassy	6	white, pink, gray	white
Mica	glassy	2 to 3	colorless, light tint	white
Quartz	glassy	7	colorless, white, pink, smoky, purple	white
Hornblende	glassy	5 to 6	black or dark green	gray-green or gray-brown

Show What You Know

1. Use the chart of mineral properties to describe the properties of calcite.

____

____

2. How could you determine if a mineral is pyrite or gold?

____

____

LESSON 30

Soil

What is soil made of?

Soil is a mixture of nonliving and once-living things. Nonliving things include water, air, small pieces of rocks, and minerals. Once-living things include matter from dead plants and animals.

Soil is made of layers. The top layer, or **topsoil**, is often rich and dark. Most plants grow in this layer. The second layer is called **subsoil**. It has less plant and animal matter, so it is lighter than topsoil and has fewer nutrients. Beneath this layer is weathered rock. The soil here is made of large pieces of rock and has little plant and animal matter. The bottom layer is made of solid rock.

Soils can be classified according to their texture. **Soil texture** depends on the size of the materials that make up the soil. Grains of clay are very tiny. Grains of silt are larger, and grains of sand are larger still. Texture affects a soil's ability to hold water and support plant growth. For example, clay has a slick, heavy texture. Water doesn't drain easily. It settles in spaces between the grains of clay, leaving no room for air. Plants that grow in these soils may die because they have too much water and not enough air. Sandy soils have a coarse texture. Water drains quickly. Plants that grow in sandy soil may dry up and die. Silt has a fine texture. Silt can hold both water and air. But the best soils for growing plants are loam soils—mixtures of silt, sand, and clay. They have loose, crumbly textures and contain plenty of water and air.

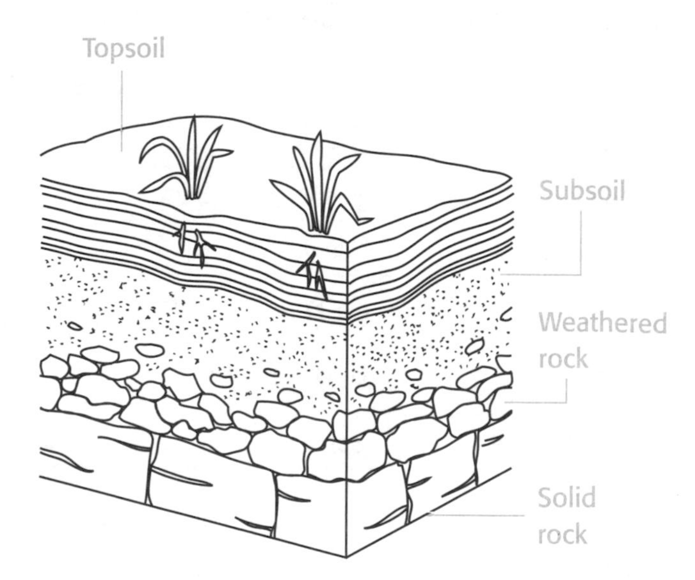

Show What You Know

Draw lines to connect each soil type to its ability to support plant life.

1. sandy soil	a. good
2. clay soil	b. best
3. silt soil	c. not enough water
4. loam soil	d. not enough air

Weathering

How does weathering affect rocks?

Since it formed billions of years ago, Earth's surface has continued to change. Some of these changes happen slowly. Weathering, for example, breaks down and changes rocks over many, many years. In time, it can make tall mountains much smaller.

Physical weathering occurs when rocks break apart. Ice can cause this type of weathering. How? Water enters cracks in rocks, then freezes and expands. The ice pushes against the rock. This can happen again and again as the ice melts and refreezes. Eventually, the rock falls apart.

Plants can also cause physical weathering. Their roots grow inside cracks in rocks, making the cracks bigger.

Chemical weathering occurs when the chemical makeup of a rock changes. Oxygen, for example, can react with water and minerals in a rock. The rock takes on a rusty look and is easier to break down. Water can also react with certain chemicals in the air to form **acid rain.** Acid rain can weather rocks and historic monuments. It also damages forests and lakes.

Physical Weathering Caused by Plants

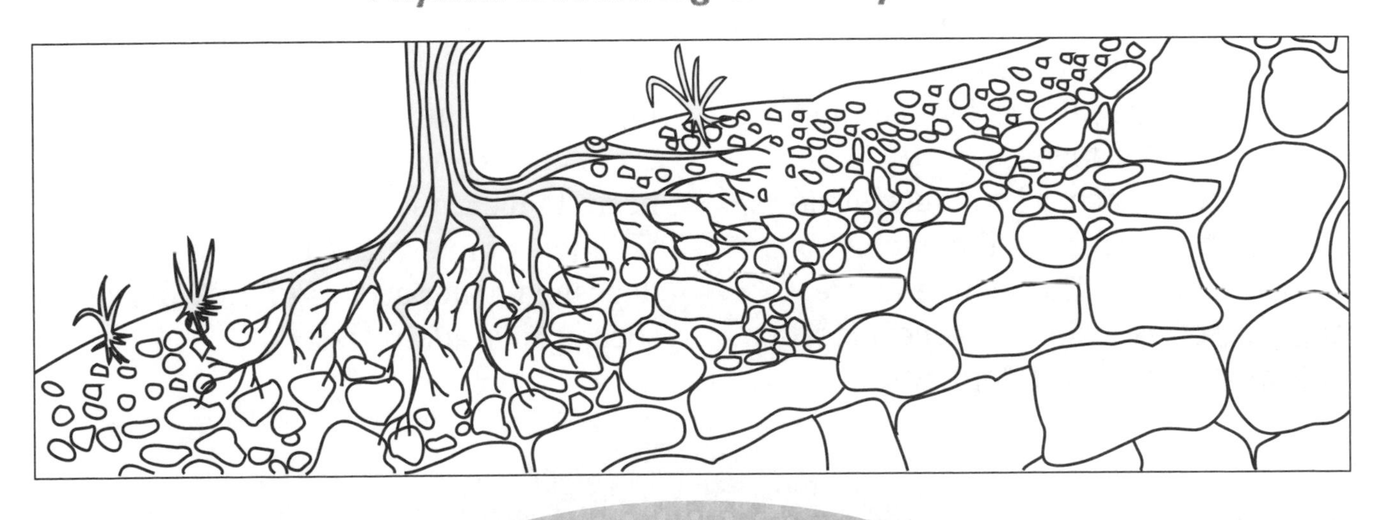

Show What You Know

Draw pictures to show how ice can physically weather a rock.

LESSON 32

Erosion

How does erosion affect Earth's surface?

Weathering and erosion work together to change Earth's surface. Weathering breaks down rocks into sediments. **Erosion** carries the sediments away. Water, wind, and glaciers can cause erosion.

Flowing water in rivers and streams can carve a path through land. The water erodes and picks up weathered sediments. This process formed the mighty Grand Canyon.

As a river reaches its mouth, it slows down. It no longer has the energy to carry sediments. The sediments are dropped, or deposited, in a new location. This process is called **deposition.** Heavy rains and ocean currents also carry and deposit sediments.

Wind picks up small sediments such as sand as it blows. Strong winds cause sediments to strike against rocks or other materials and weather them.

Glaciers are large masses of ice and snow that move slowly on land. Working like huge shovels, they carry eroded material in front of them. Glaciers can carve out valleys or flatten hills. As they begin to melt, they deposit their sediments in new locations. The deposited materials form new features such as hills and ridges.

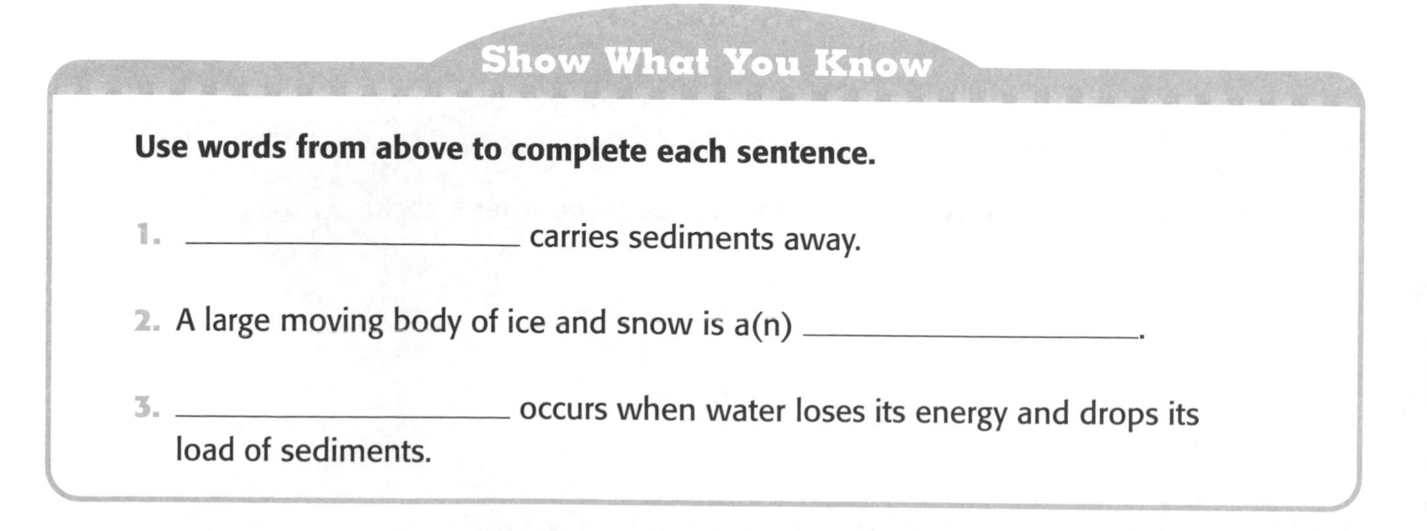

Show What You Know

Use words from above to complete each sentence.

1. ____________________ carries sediments away.

2. A large moving body of ice and snow is a(n) ____________________.

3. ____________________ occurs when water loses its energy and drops its load of sediments.

Floods and Landslides

What causes floods and landslides?

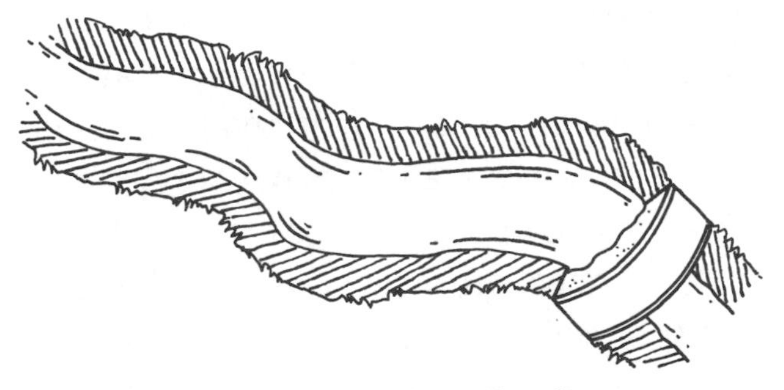

Levees and dams help reduce flooding.

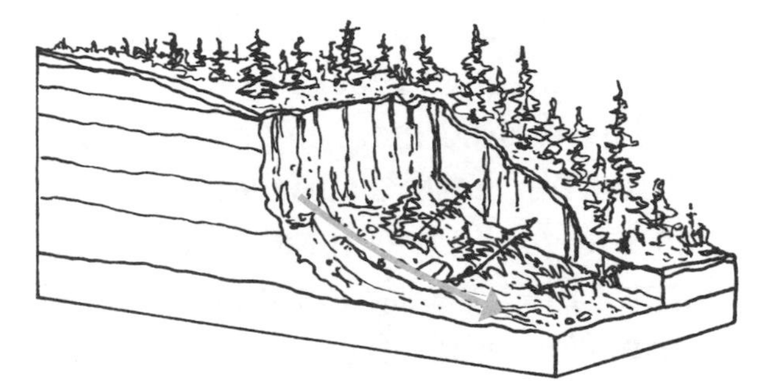

A landslide occurs when soil becomes wet and slides downward.

Some changes on Earth occur very rapidly. **Floods,** for example, occur when rivers and streams overflow their banks. Floods are caused by heavy rains or fast melting snow. The most powerful floods can uproot trees and wash away bridges and homes. Each year, floods cause billions of dollars of damage in the United States alone. People build **levees** and **dams** to help reduce flooding.

Landslides also cause rapid changes to Earth's surface. Landslides occur when **gravity** moves large amounts of soil and other materials down a slope. Some landslides happen after heavy rains. The soil becomes too wet and slides downward. Other landslides happen when rocks on a steep slope tumble down.

Landslides can destroy homes and other structures in their path. Certain human activities, such as construction, can increase landslides. For this reason, people are often discouraged from building on steep slopes. Planting vegetation is one way to prevent landslides. The roots of the plants help hold the soil together.

Show What You Know

A company wants to build apartments on a steep slope. Do you think this is a good idea? Write a brief paragraph explaining your views.

____

____

____

____

LESSON 34

Volcanoes and Earthquakes

What are volcanoes and earthquakes?

Magma deep beneath Earth's **crust** can flow out through cracks in the surface. The melted material, called **lava** when it reaches Earth's surface, can build up and cool. Over time, a mountain may form. This type of mountain is called a **volcano.** Volcanoes erupt gases and ash, along with lava.

There are three different types of volcanoes. **Shield volcanoes** have quiet eruptions. Their slopes are gentle. **Cinder cone volcanoes** have violent eruptions. They have steep slopes. **Composite volcanoes** have both quiet and violent eruptions. Their slopes are made of layers of lava over layers of rock.

Earthquakes are waves of energy caused by movements of rock. Beneath Earth's surface, great sections of rock push against or slide past each other. Intense pressure causes the rocks to break. The rocks move suddenly along an area called a **fault.** This movement releases waves of energy. Sometimes, the waves are barely felt. Other times, whole cities are destroyed as the land drops and tilts.

Volcano Types

Shield volcano

Cinder cone volcano

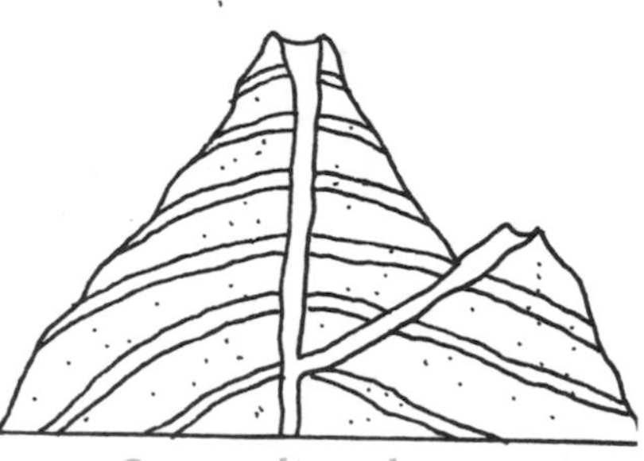

Composite volcano

Show What You Know

List the following stages of an earthquake in the order they happen.

a. Energy is released.

b. Rocks break.

c. Rocks move suddenly along a fault.

d. Rocks push against each other.

____ ____ ____ ____

A Multiple Choice

Fill in the letter to show your answer.

1. **Which of the following processes cause a sedimentary rock to change into a metamorphic rock?**

   Ⓐ weathering and erosion

   Ⓑ heat and pressure

   Ⓒ cooling and hardening

   Ⓓ melting

2. **Which type of rock is made of hardened magma?**

   Ⓐ igneous

   Ⓑ sedimentary

   Ⓒ metamorphic

   Ⓓ fossil

3. **Which statement about the rock cycle is true?**

   Ⓐ Rocks never change.

   Ⓑ Heat and pressure form igneous rocks.

   Ⓒ Rocks can only change from igneous to sedimentary.

   Ⓓ Rocks constantly change into other rocks.

4. **In which soil layer do most plants grow?**

   Ⓐ topsoil

   Ⓑ subsoil

   Ⓒ weathered rock

   Ⓓ solid rock

5. Which statement about cinder cone volcanoes is true?

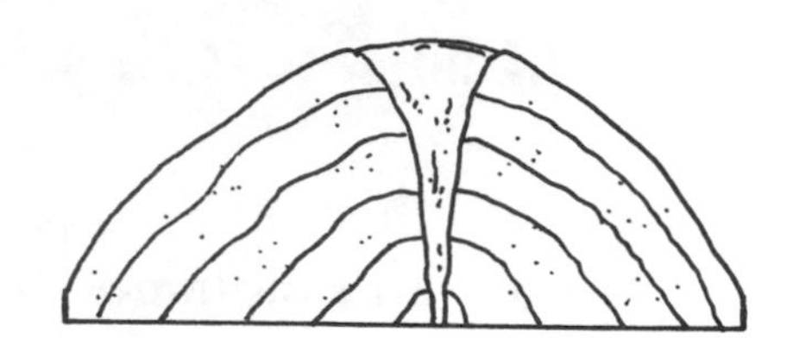

Ⓐ It has steep slopes made by violent eruptions.

Ⓑ It has gentle slopes made by violent eruptions.

Ⓒ It has gentle slopes made by quiet eruptions.

Ⓓ It has gentle slopes made by both quiet and violent eruptions.

6. Why do wind and water deposit sediments?

Ⓐ They pick up sediments that are too large.

Ⓑ They lose energy.

Ⓒ They can only carry sediments for short distances.

Ⓓ They gain speed.

Short Response

7. The roots of a plant are growing inside the cracks in a rock. What will likely happen to the rock?

____

____

____

____

____

____

Earth's Water

Where is water found on Earth?

Earth is often called the "blue planet." Why? **Water** covers nearly 75 percent of Earth's surface. Most of this water is salty. It is found in oceans. Only about 3 percent of Earth's water is fresh, or not salty.

About 90 percent of freshwater on Earth is locked up in glaciers and ice caps near the north and south poles. People cannot use this water for drinking or other needs.

People can use freshwater found in streams, rivers, ponds, lakes, and underground. Water that soaks underground is called **groundwater.** Groundwater collects in spaces between rocks. People can drill wells to reach the water.

On Earth, water continually moves from the **atmosphere** to the surface and back again. This process is known as the **water cycle.** Energy from the sun drives the water cycle.

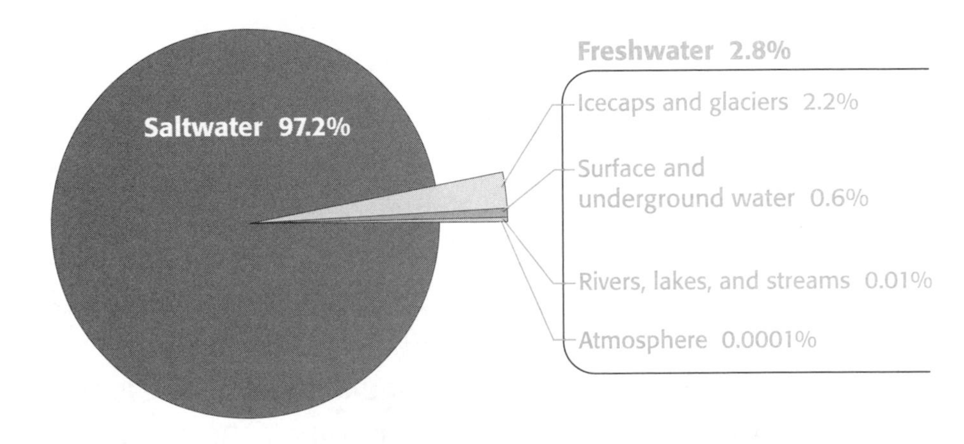

Show What You Know

1. How is lake water different than ocean water?

____

____

2. If water covers most of Earth's surface, why do we need to conserve water?

____

____

LESSON 36

The Water Cycle

What processes make up the water cycle?

The three processes that make up the water cycle are evaporation, condensation, and precipitation.

In the process of **evaporation,** heat energy from the sun causes water to change from a liquid to a gas. This gas is called **water vapor.** It rises from the surfaces of land, oceans, rivers, and lakes and enters the atmosphere.

During **condensation,** water vapor in the atmosphere cools and changes back to tiny droplets of liquid water. Millions of tiny droplets join together to form clouds.

Droplets continue to combine until they become too large to be held up by air. In the process of **precipitation,** the water falls back to Earth's surface. Precipitation can be in the form of rain, snow, sleet, hail, or drizzle.

Precipitation can fall directly onto bodies of water. Energy from the sun then evaporates the water and the cycle starts again. Precipitation can also fall on the ground. There, it may soak beneath the soil and become groundwater. Or, it may become runoff. **Runoff** is water that runs along Earth's surface. Some runoff evaporates immediately. Some reaches rivers, lakes, and oceans.

Show What You Know

What role does heat energy play in the different stages of the water cycle?

____

____

____

Weather

What is weather?

Earth's atmosphere is made of different layers. The layer closest to Earth's surface is called the **troposphere.** Storms and other types of weather take place in the troposphere. **Weather** describes the state of the atmosphere at a given time and place.

Different conditions affect the type of weather in an area. **Temperature,** for example, describes how warm or cool the air is. Air is made of **molecules.** When molecules move quickly, the air feels warm. When molecules move slowly, the air feels cool. Temperature changes throughout the day as amounts of energy from the sun change. For the same reason, temperature also varies from month to month. **Thermometers** measure temperature.

As the molecules that make up air move, they push in all directions. The force of air pushing is called **air pressure.** Air pressure affects weather. Rising air pressure is generally a sign of clear weather to come. Falling air pressure is a sign of stormy weather. **Barometers** measure air pressure.

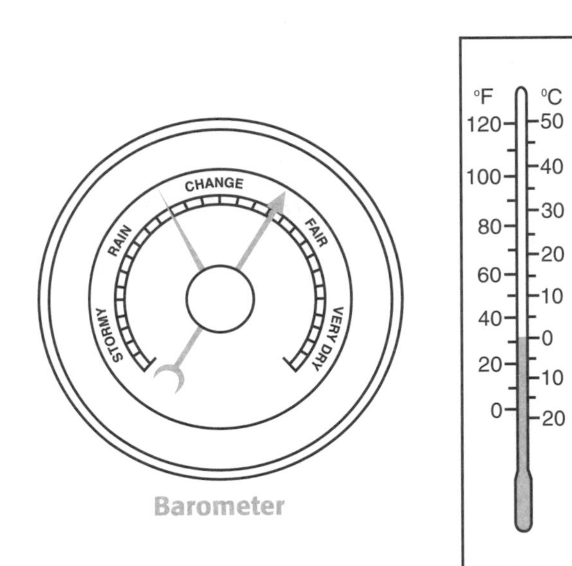

Barometer

Thermometer

Show What You Know

Use the terms below to make a concept map about temperature.

temperature **air** **molecules** **move** **quickly** **slowly** **warm** **cool**

LESSON 38

Weather

What other factors affect weather?

Weather is also affected by wind. **Wind** is the movement of air. Air moves because the sun does not heat Earth's surface evenly. This causes differences in air pressure. In general, wind moves from areas of high pressure to areas of low pressure. **Wind vanes** show wind direction. **Anemometers** measure wind speed.

Humidity is the amount of water vapor in the air. **Relative humidity** is the amount of water vapor in air compared to the amount the air could hold at a given temperature. When the air cannot hold any more water vapor, its relative humidity is 100 percent. Air can hold more water vapor at warm temperatures than it can at cool temperatures. **Hygrometers** measure relative humidity.

Relative Humidity Changes in Relation to Temperature

Air Temperature (°C)	-20°	-10°	0°	10°	20°	30°	40°
Water Vapor (g/m³)	2	3	5	8	15	30	45

As its temperature rises, air can hold more water vapor.

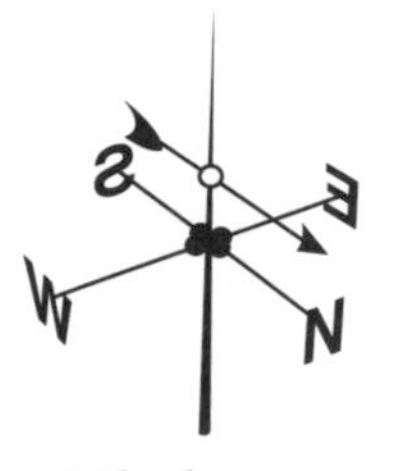

Wind vane

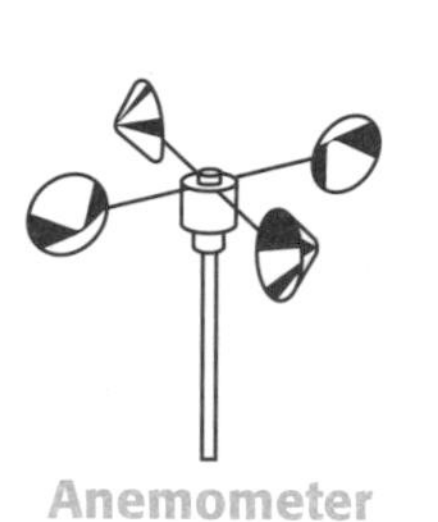

Anemometer

Hygrometer

Show What You Know

Imagine that the air temperature in your classroom increased by 10°C. Would the relative humidity of the air in the room go up or down? Explain.

____

____

____

Clouds

How are clouds classified?

Clouds are made of billions of tiny droplets of water. They form when water vapor in the air condenses, or changes state from a gas to a liquid. Clouds are classified based on their shape and the height at which they form.

There are three main types of cloud shapes. **Stratus** clouds form flat layers or sheets across the sky. They often bring precipitation. **Cumulus** clouds are puffy and white. They can bring fair or stormy weather. **Cirrus** clouds are feathery. They usually signal fair weather ahead.

Clouds form at different heights in the atmosphere. For example, the word part **"cirro-"** describes clouds that form at very high heights. **"Alto-"** describes clouds that form at middle heights. **"Strato-"** describes low clouds. Weather forecasters often combine these word parts with words that describe cloud shapes. A cirrostratus cloud, for example, looks like a flat sheet high in the sky.

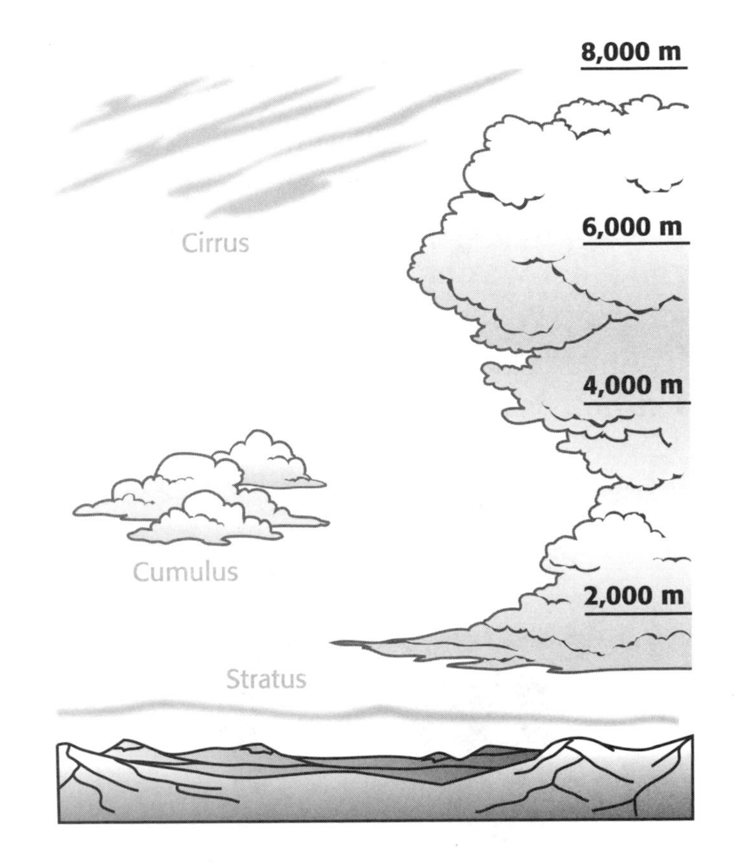

Another word part, **"nimbus,"** describes clouds that produce precipitation. Precipitation can fall in the form of rain, snow, hail, sleet, or drizzle.

Show What You Know

Where in the sky would a nimbostratus cloud be found? What kind of weather would it bring? Explain your answer.

____

____

____

____

LESSON 40

Climate

What factors affect climate?

Recall that weather describes what is happening in the atmosphere at a certain time. **Climate** describes the long-term weather in an area. Scientists study patterns of precipitation, temperature, and other weather factors to determine an area's climate.

An area's climate is affected by latitude. **Latitude** is the distance north or south of the equator. **Tropical** climates are close to the equator. They get the most direct sunlight and are warm year-round. **Temperate** climates lie between tropical and polar regions. They have moderate temperatures. **Polar** climates are near the poles. They get little direct sunlight and are cold year-round.

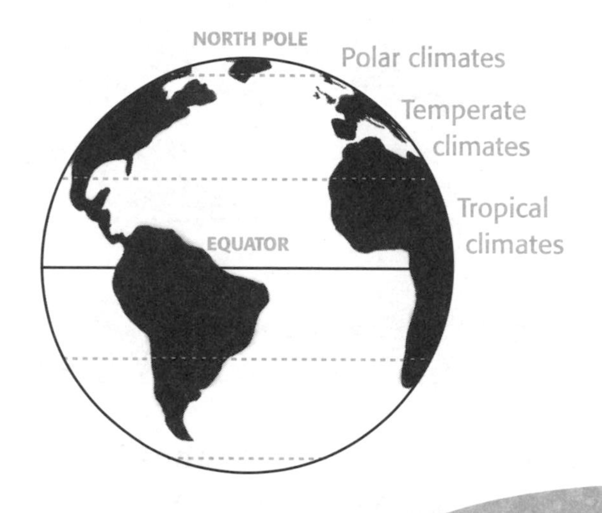

Large bodies of water also affect climate. Water heats up and cools down more slowly than land does. Thus, places that are close to oceans or lakes tend to have warmer winters and cooler summers than places farther inland.

Elevation describes the height of an area above sea level. In general, places with higher elevations have cooler climates than places with lower elevations.

Show What You Know

Temperatures in Swiss Cities of Similar Latitude

City	Average Yearly Temperature (°F)
Bern	49°
Bignasco	32°
Gams	30°
Lugano	53°
Zermatt	41°

Which city in the table do you think is located at the highest elevation? Explain your answer.

____

Multiple Choice

Fill in the letter to show your answer.

1. **Most water on Earth is found in**

   Ⓐ rivers.

   Ⓑ oceans.

   Ⓒ ice caps.

   Ⓓ lakes.

2. **What is the source of energy for the water cycle?**

   Ⓐ water

   Ⓑ air pressure

   Ⓒ the sun

   Ⓓ Earth's atmosphere

3. **Choose the statement that best describes the water cycle.**

   Ⓐ During evaporation, water freezes.

   Ⓑ During condensation, water vapor enters the atmosphere.

   Ⓒ Precipitation forms clouds.

   Ⓓ During condensation, water changes from a gas to a liquid.

4. **If a weather forecaster wanted to measure the temperature of air, she would use a(n)**

   Ⓐ thermometer.

   Ⓑ barometer.

   Ⓒ anemometer.

   Ⓓ hydrometer.

5. You would like to live in a climate that is warm year-round. Where should you live?

Ⓐ near the poles

Ⓑ in a temperate region

Ⓒ near the equator

Ⓓ on a mountain top

6. Which type of cloud is shown here?

Ⓐ cirrus

Ⓑ stratus

Ⓒ cumulus

Ⓓ nimbus

Short Response

7. Jose measured the relative humidity outside his classroom several times in one day and recorded his measurements in this chart. When was the air unable to hold more water vapor? Explain your answer.

Relative Humidity

Time	Measurement
8 A.M.	75 percent
10 A.M.	87 percent
12 P.M.	94 percent
2 P.M.	100 percent

____

____

____

____

Resources

What are resources?

Resources are the things people take from the environment and use to live. They include soil, water, air, rocks, minerals, plants, and animals.

Humans breathe air and drink water. They use soil and water to grow plants. They use plants and animals for food, clothing, and shelter. They even use chemicals found in plants and fungi to make medicines.

In many parts of the world, humans use rocks and minerals to build homes, offices, and other buildings. They use glass to form walls and windows and minerals such as copper to make electrical wires.

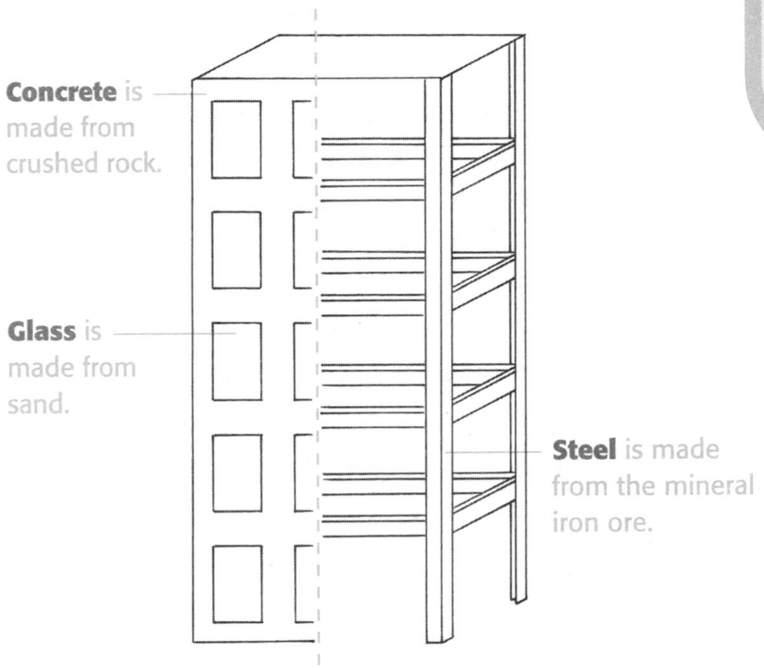

A building made with different resources

Show What You Know

Give one example of how each resource listed below is useful to people.

1. air ______________________________

2. animals ______________________________

3. minerals ______________________________

4. plants ______________________________

5. rocks ______________________________

6. soil ______________________________

7. water ______________________________

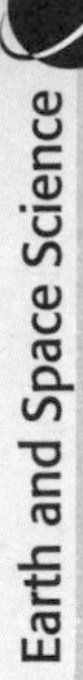

LESSON 42

Conserving Resources

How can you conserve resources?

Some resources are **renewable.** That means they can be replaced as they are used. For example, if a tree is cut down, a new tree can be planted in its place. Other resources are **nonrenewable.** People use them far faster than they can be replaced. Minerals are an example of a nonrenewable resource.

Conservation is the careful and wise use of all resources. Such careful use makes sure there are enough resources for people to use today and in the future.

One way people can make resources last longer is to **reduce** the amount of resources they use. Taking cloth bags to the grocery store reduces the number of trees needed to make paper bags. Fixing a leaky faucet reduces water use. Repairing a broken garden tool rather than throwing it away reduces the need for new minerals.

Reusing items also helps conserve resources. You can reuse paper bags instead of buying new bags. A **compost pile** is a great way to turn food that would have been thrown away into something valuable—rich soil for growing plants.

Many used resources can be **recycled.** Glass, newspapers, metal cans, and plastic containers, for example, can be re-made into the same or new materials. Items that show the recycling symbol can be recycled.

Recycling symbol

Show What You Know

List one way you could reduce, reuse, and recycle a resource every day.

1. Reduce: ______________________

2. Reuse: ______________________

3. Recycle: ______________________

LESSON 43

Fossil Fuels

What are fossil fuels?

Humans use energy to do lots of things. We use energy to heat our homes, light up our classrooms, and run our cars, buses, and factories. We get the energy we need from **energy resources.**

Energy resources can be either renewable or nonrenewable. **Renewable energy sources**, like wind or solar energy, do not run out. **Nonrenewable energy sources** are limited. They cannot be replaced once they are used up.

The most commonly used energy resources are **fossil fuels,** which include coal, oil, and natural gas. Coal is burned to produce electricity. Oil is used to make gasoline. Natural gas is used to heat homes.

Fossil fuels form from the remains of plants and animals that died millions of years ago. The decayed material is squeezed together and, over time, changes into a fossil fuel.

Fossil fuels are a nonrenewable resource. That's because we use fossil fuels much more quickly than they are formed.

Formation of Fossil Fuels

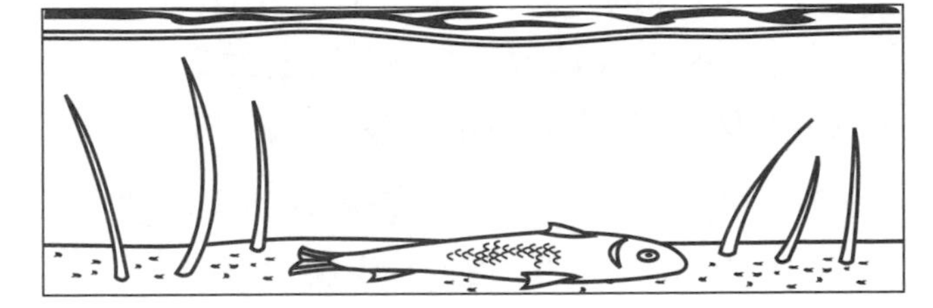

A Ancient plants and animals died and settled to the bottom of the ocean or lake where they lived.

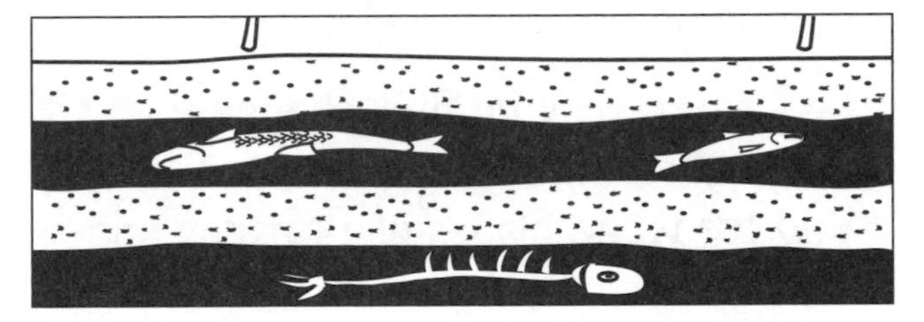

B Sand and mud buried the remains. Layers of remains, sand, and mud built up.

C Over millions of years, heat and pressure turned the remains into oil, coal, and natural gas.

Show What You Know

List three ways you use energy from fossil fuels every day.

1. ______________________________

2. ______________________________

3. ______________________________

LESSON 44

Renewable Energy Resources

What are some sources of renewable energy?

People everywhere depend on energy to live. As supplies of nonrenewable fossil fuels run lower, people need renewable energy sources.

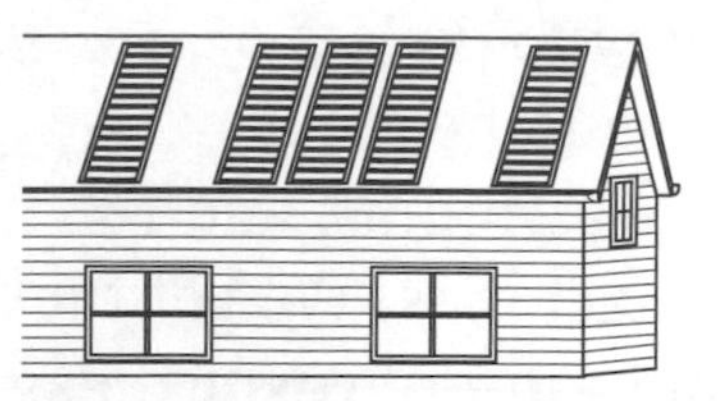
Solar cells on a roof

Hydroelectric Power The energy of moving water can be used to make electricity. Dams built across rivers force moving water through the blades of a **turbine.** A generator turns the energy of the moving blades into electricity.

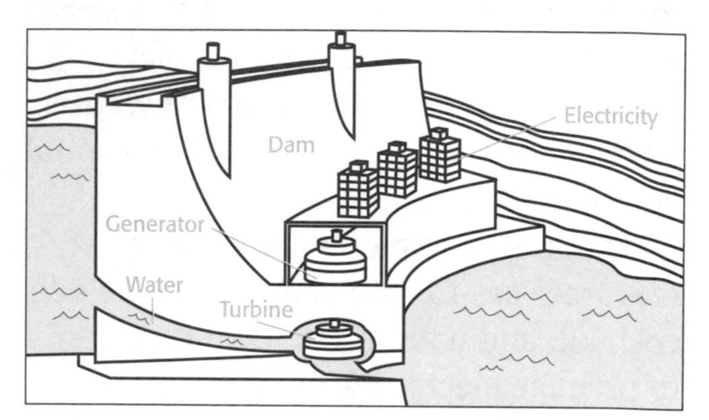

Hydroelectric dam

Wind Power Windmills change the energy of moving air into electricity. Wind turns the blades of a turbine and creates electricity.

Solar Energy Energy from the sun is another source of renewable energy. **Solar cells** absorb energy from the sun and change it to electricity.

Geothermal Energy Heat energy beneath Earth's surface can also be used to make electricity. Water is pumped deep underground. There, magma heats the water and turns it into steam. The steam rises up through pipes to a power plant on the surface. The steam turns the blades of a turbine and creates electricity.

Show What You Know

List four types of renewable energy and describe where they come from.

Renewable Energy Resource	Where Energy Comes From
1. ______________________	______________________
2. ______________________	______________________
3. ______________________	______________________
4. ______________________	______________________

Water Pollution

How does pollution hurt our water resources?

Human activities can change air, soil, and water in unwanted ways. These harmful changes are called **pollution.** There is a limited supply of freshwater on Earth that all living things must share. Even a small amount of pollution can ruin a large water source.

Factories sometimes release chemicals into water. The pollution travels downstream.

Water pollution can come from many different sources. Chemical fertilizers used on many farms and golf courses can wash into nearby water sources. Factories sometimes dump chemicals and waste into lakes and streams. Factories may also release heated water that can harm or kill fish.

Human waste, or **sewage,** can also pollute water. If sewage is dumped into a lake or stream without being treated, it can kill organisms that live in the water. It can also make someone who uses the polluted water sick. Chemicals from buried garbage can seep into groundwater. They can also harm or kill organisms that use the water.

Nature can pollute water, too. For example, wind can drop lots of soil and sand into water. The soil and sand keep sunlight from reaching the plants and animals that live in the water. This harms or kills the organisms.

Show What You Know

Explain where each type of pollution comes from.

1. Chemicals ______________________________
2. Heat ______________________________
3. Sewage ______________________________
4. Garbage ______________________________
5. Soil and sand ______________________________

LESSON 46

Air Pollution

How does pollution harm our air?

Most living things need air to live. But polluted air can be dangerous to plants, animals, and even buildings. Air pollution is caused by harmful substances being released into the air. Usually, these harmful substances come from human activities, like burning fossil fuels. But not all air pollution is caused by humans. Forest fires produce smoke that pollutes the air. Big storms can send large amounts of dust into the air, and volcanic eruptions blow ash into the sky.

One of the biggest air pollution problems is **acid rain.** Acid rain forms when pollution from cars and factories combines with water in the air and falls as rain. When acid rain falls on forests and lakes, the acids can harm or kill the plants and animals that live there. Acid rain also damages buildings and statues by wearing away stone and brick.

Ozone is a gas in Earth's atmosphere. A thick layer of ozone around Earth blocks harmful rays from the sun. Some scientists think certain kinds of air pollution can destroy ozone. Scientists have discovered holes in the ozone layer above the north and south poles. These holes appear to be getting larger each year.

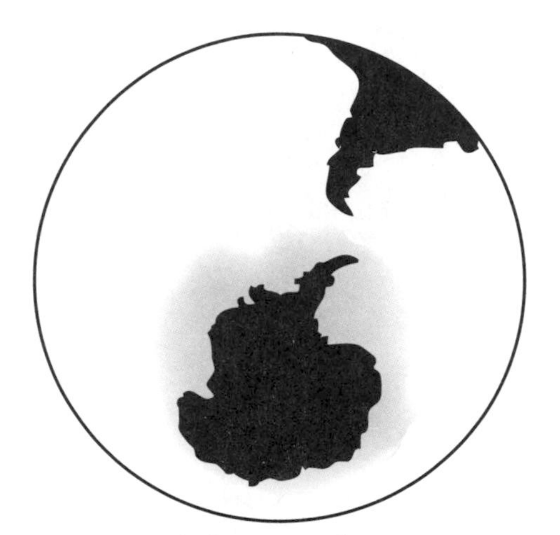

Hole in ozone layer near south pole

Show What You Know

1. Name one human activity that pollutes the air. ____________________

2. Name one way nature pollutes the air. ____________________

3. Draw a diagram to explain how acid rain forms and how it affects things on Earth.

A Multiple Choice

Fill in the letter to show your answer.

1. **Which of the following is an example of a renewable energy resource?**

   (A) oil

   (B) coal

   (C) natural gas

   (D) solar energy

2. **Which of the following do we depend on mineral resources for?**

   (A) food

   (B) fabrics

   (C) electrical wire

   (D) medicine

3. **Soil and sand pollute water by**

   (A) making people who drink it sick.

   (B) blocking light.

   (C) causing too many plants to grow.

   (D) releasing harmful chemicals.

4. **A product that shows this symbol on its packaging can be**

   (A) reused.

   (B) burned.

   (C) recycled.

   (D) returned.

• • EARTH AND SPACE SCIENCE TEST C • •

Use the diagram to answer Item 5.

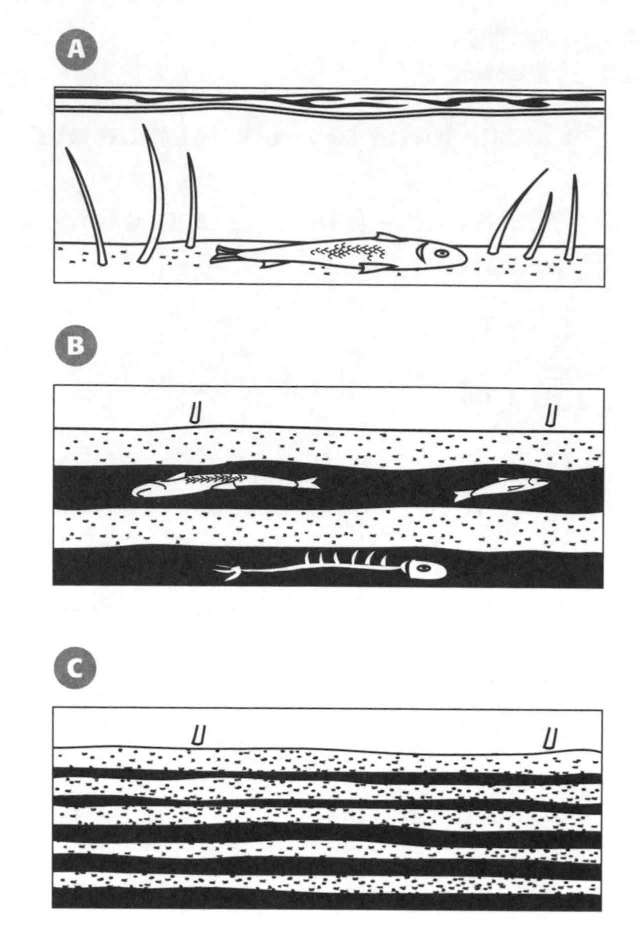

5. **How long does it take for fossil fuels to form?**

   Ⓐ several years

   Ⓑ hundreds of years

   Ⓒ thousands of years

   Ⓓ millions of years

6. **Where does the energy in fossil fuels come from?**

   Ⓐ heat energy in Earth

   Ⓑ moving water

   Ⓒ wind

   Ⓓ remains of dead plants and animals

Short Response

7. **Why should people use renewable rather than nonrenewable energy resources?**

____

____

____

____

____

LESSON 47

The Solar System

What makes up our solar system?

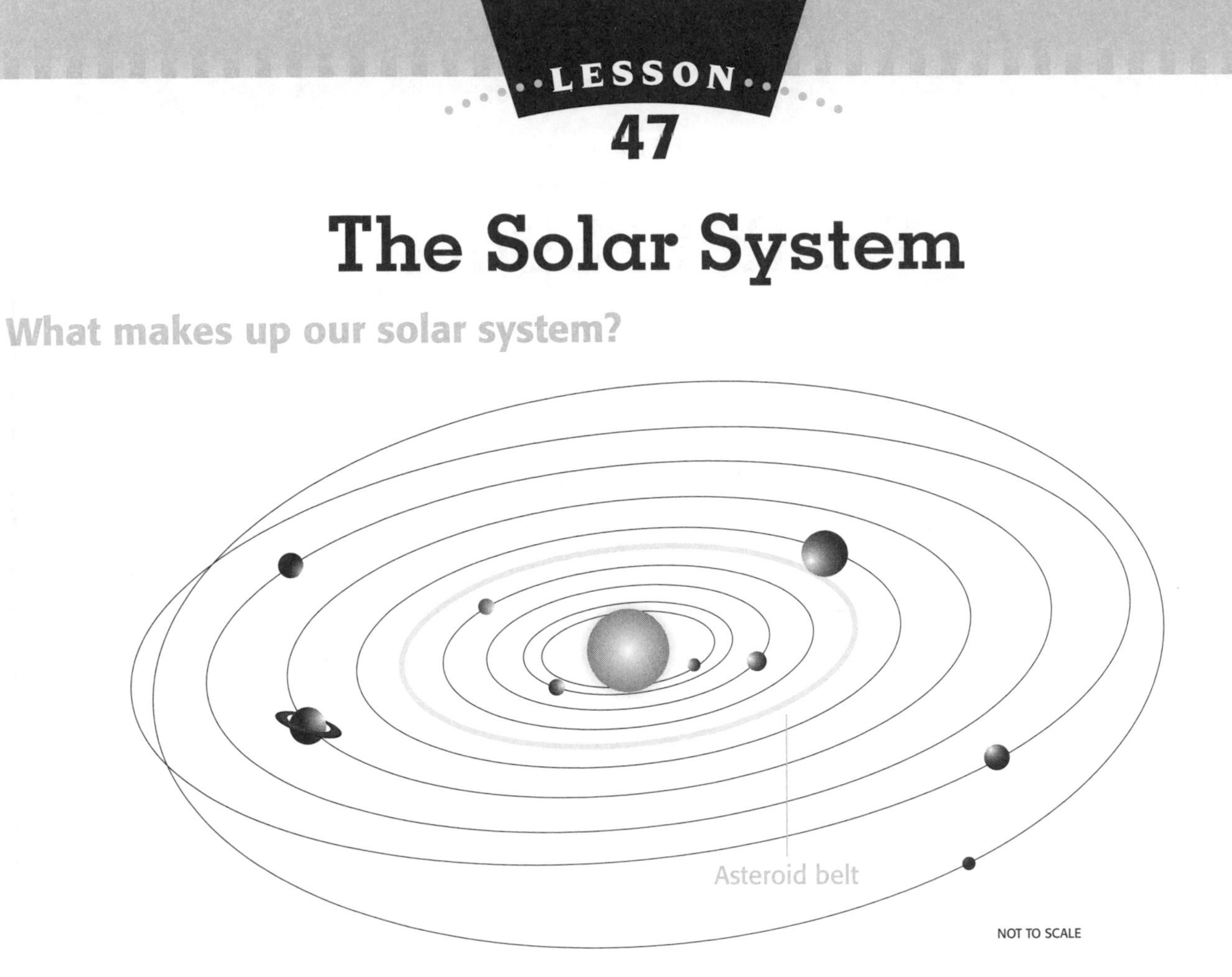

Our solar system includes the planets, their moons, and other objects that orbit the sun.

Our **solar system** includes the sun and nine planets and their moons. It also includes a number of comets and asteroids. Beyond our solar system are other stars and their solar systems.

The sun is the center of our solar system. It has the most **mass** of all the objects in the solar system. The more mass an object has, the greater its force of **gravity.** The Sun's gravity is strong enough to keep all the planets in the solar system in orbit around it. The oval-shaped path a planet follows as it **revolves** around the sun is called its **orbit.** It takes Earth about one year to revolve once around the sun.

Comets are big pieces of ice that orbit the sun. We see comets as streaks of light with a bright tail. **Asteroids** are large rocks that orbit the sun.

Show What You Know

List all the objects that are part of our solar system.

____

____

LESSON 48

Earth in Space

What causes night and day and the seasons?

As Earth revolves around the sun, it also **rotates** on its **axis.** It takes Earth one day, or 24 hours, to rotate once around. As it rotates, the side of Earth that faces the sun experiences daytime. The side that faces away from the sun experiences nighttime.

Earth is tilted on its axis. This tilt causes the sun's rays to strike some places on Earth more directly during certain times of the year. This causes the **seasons.**

In July, August, and September, it is summer in the Northern Hemisphere and winter in the Southern Hemisphere. That is because the Northern Hemisphere is tilted toward the sun, so it receives more direct sunlight. In January, February, and March, the Southern Hemisphere receives more direct sunlight. So during those months, it is summer in the Southern Hemisphere and winter in the Northern Hemisphere.

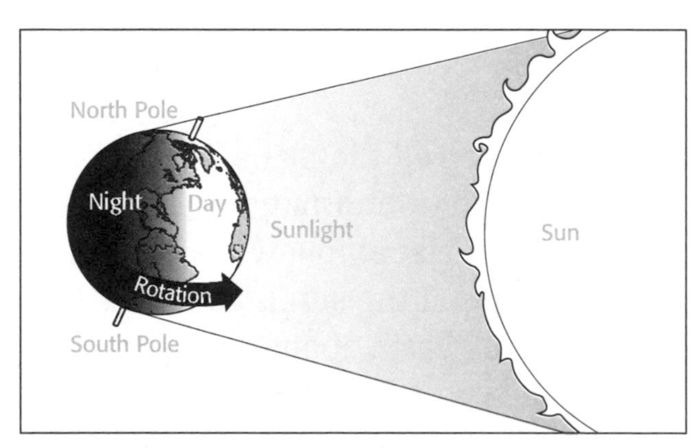

Earth rotates from west to east. That's why the sun appears to rise in the east and set in the west.

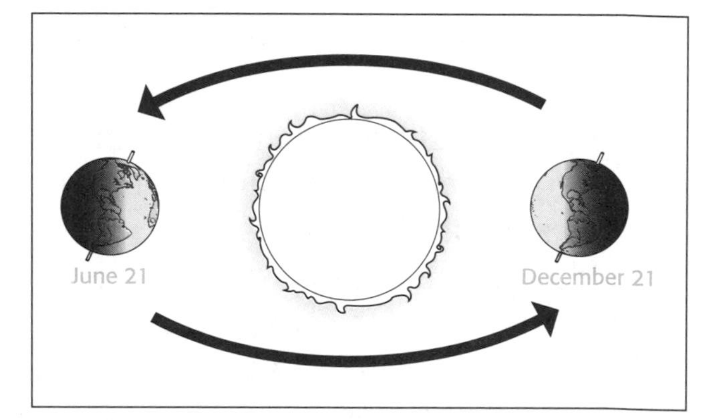

The Earth's tilt causes the seasons.

Show What You Know

Explain why there would be no seasons if Earth were not tilted on its axis.

____

____

____

Moon Phases

Why does the moon change shape?

From Earth, the moon looks like it has a different shape from night to night. These different shapes are the **phases** of the moon. The moon doesn't actually change shape. What changes is the part of the moon that is reflecting light from the sun. And that depends on where the moon is in its orbit around Earth.

Occasionally, the moon moves through Earth's shadow. From Earth, it looks like the moon has disappeared. This is a **lunar eclipse.** An eclipse lasts until the moon moves out of Earth's shadow.

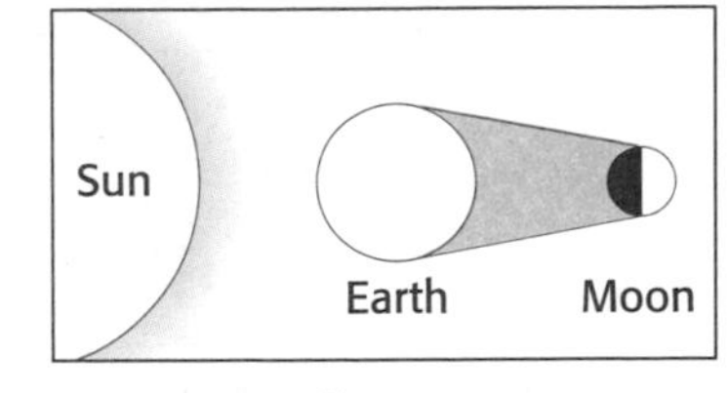

A Lunar eclipse

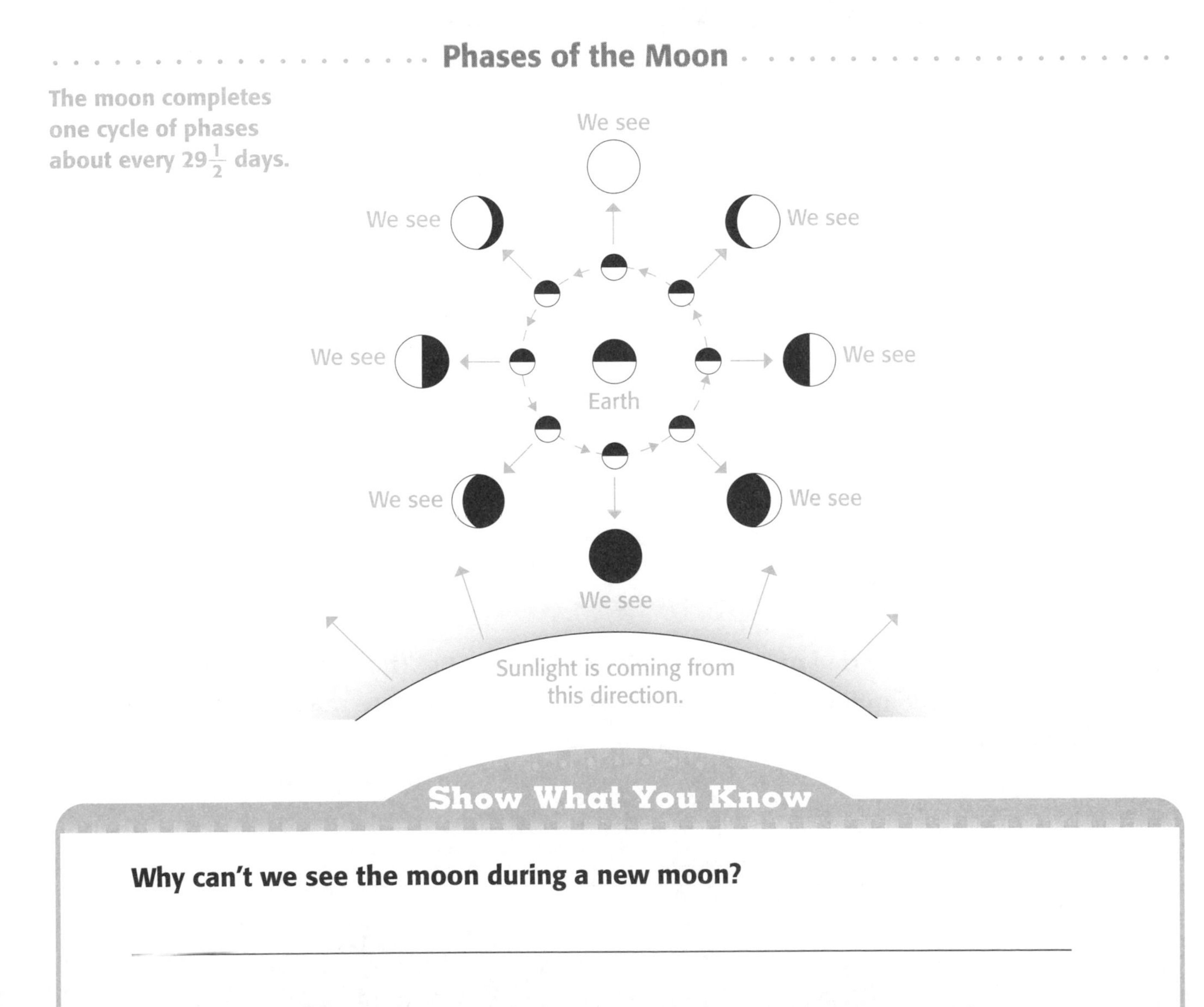

Show What You Know

Why can't we see the moon during a new moon?

____

____

LESSON 50

The Moon's Surface

What is the surface of the moon like?

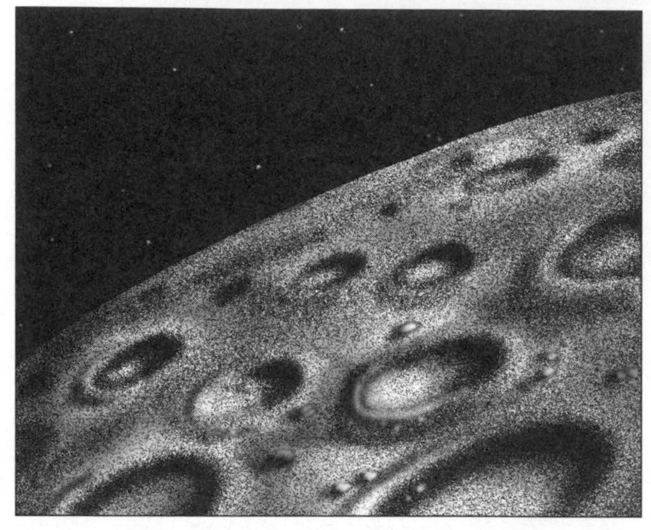
The moon's surface is covered with craters.

Many of the planets in our solar system have moons. Earth's moon is one of the largest. The surface of the moon is covered in craters. These craters were formed by rocks from space that hit the surface.

The moon's surface is rocky and dry. There is no water or **atmosphere.** Because there is no atmosphere, there is no weather. As a result, the moon's surface is not worn away by storms and moving water as Earth's surface is. Since the surface does not change, the craters look the same today as they did when they were made.

Earth's atmosphere keeps it from heating up too much or too quickly when the sun is shining on it. The atmosphere also keeps it from cooling down too much or too quickly at night. Because the moon has no atmosphere, temperatures on the surface quickly go from very cold to very hot. In full sunlight, the surface can reach temperatures of over 120°C. When it is dark, the temperatures can drop to below -170°C.

Show What You Know

Astronauts first landed on the moon in the late 1960s. Their footprints on the surface still look exactly the same as the day they were made. Why is that?

____

____

____

____

The Planets

What planets are in our solar system?

There are nine planets in our solar system. Scientists divide the planets into four inner planets and five outer planets. The inner planets are smaller than most of the outer planets and are made of dense rock. The outer planets, except for Pluto, are very large and made of gases.

The inner planets, in order of closeness to the sun, are Mercury, Venus, Earth, and Mars.

The outer planets, in order of closeness to the sun, are Jupiter, Saturn, Uranus, Neptune, and Pluto.

	Planet	Orbit Time	Notes
Inner Planets	**Mercury**	88 days	Very small, almost no atmosphere
	Venus	225 days	Dense atmosphere, very hot
	Earth	365 days	Has atmosphere, water, supports life
	Mars	687 days	Dry, no life, but signs that water was once present
Outer Planets	**Jupiter**	about 12 years	The largest planet; thick, stormy atmosphere
	Saturn	about 30 years	Has bright rings made of ice and rock
	Uranus	about 84 years	Blue-green ball of gas and liquid
	Neptune	about 165 years	Surface of rock and frozen water; thick atmosphere
	Pluto	about 249 years	Very small and rocky

Show What You Know

Think of a saying that will help you remember the order of the planets in our solar system.

___

___

___

LESSON 52

The Universe

What makes up the universe?

A **star** is an object in space that makes its own heat and light. Stars come in a variety of sizes, colors, and brightness levels. Our sun is a medium-sized star. Like all stars, it is a burning ball of gases.

A **galaxy** is a group of millions or even billions of stars. Galaxies have different sizes and shapes. Our solar system is part of the Milky Way galaxy. Our sun is just one of billions of stars in the Milky Way galaxy.

The Milky Way galaxy is a spiral galaxy. It is shaped like a disk with spiral arms. Our sun is located near the edge of the disk. The Milky Way galaxy is just one of several billion in the universe. All the galaxies together make up the **universe.**

The Milky Way Galaxy

Our sun is a star located in the Milky Way Galaxy.

Show What You Know

Make a concept map showing how the terms *sun, star, solar system, galaxy,* and *universe* are related.

• • EARTH AND SPACE SCIENCE TEST D • •

A Multiple Choice

Fill in the letter to show your answer.

1. **All the planets in the solar system orbit the**

   Ⓐ sun.

   Ⓑ Earth.

   Ⓒ moon.

   Ⓓ stars.

2. **The reason that Earth's day is 24 hours long is because**

   Ⓐ it takes Earth 24 hours to orbit the sun.

   Ⓑ it takes Earth 24 hours to make one rotation on its axis.

   Ⓒ Earth tilts toward and away from the sun every 24 hours.

   Ⓓ the moon moves between Earth and the sun every 12 hours.

3. **In this picture, the Southern Hemisphere of Earth is experiencing**

   Ⓐ summer.

   Ⓑ fall.

   Ⓒ winter.

   Ⓓ spring.

4. **The sun rises in the east because**

   Ⓐ Earth revolves from west to east.

   Ⓑ The sun revolves from east to west.

   Ⓒ The sun rotates from east to west.

   Ⓓ Earth rotates from west to east.

5. **Which planet is closest to the Sun?**

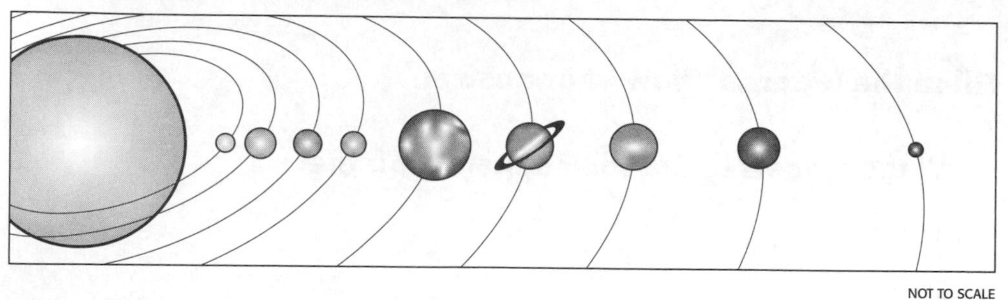

NOT TO SCALE

Ⓐ Mars

Ⓑ Earth

Ⓒ Mercury

Ⓓ Pluto

6. **Which of the following outer planets is made of solid rock?**

Ⓐ Jupiter

Ⓑ Neptune

Ⓒ Pluto

Ⓓ Saturn

7. **Which planet takes the longest time to orbit the sun?**

Ⓐ Mars

Ⓑ Pluto

Ⓒ Venus

Ⓓ Saturn

Short Response

8. **What is the difference between a solar system and a galaxy?**

___

___

___

Check your answers to questions on Earth and Space Science Test A on pages 51-52.

1. **Which of the following processes cause a sedimentary rock to change into a metamorphic rock?**

   (B) heat and pressure

   Heat and pressure form metamorphic rocks.

2. **Which type of rock is made of hardened magma?**

   (A) igneous

   Igneous rocks form from magma or lava. The magma cools, hardening into rock.

3. **Which statement about the rock cycle is true?**

   (D) Rocks constantly change into other rocks.

   In the rock cycle, rocks constantly change from one form to another. They can also change back into the same type of rock.

4. **In which soil layer do most plants grow?**

   (A) topsoil

   The top layer is rich and dark. It is where most plants grow.

5. **Which statement about cinder cone volcanoes is true?**

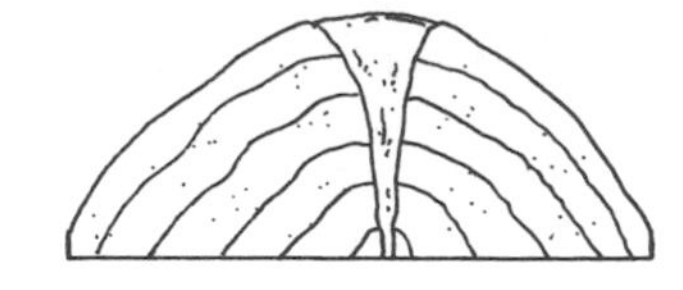

   (A) It has steep slopes made by violent eruptions.

   Violent eruptions occur over and over again. With each eruption, new lava flows downward. It cools and forms steep slopes.

6. **Why do wind and water deposit sediments?**

   (B) They lose energy.

   When wind and water lose energy, they can't carry as many sediments and drop, or deposit, them.

7. **The roots of a plant are growing inside the cracks in a rock. What will likely happen to the rock?**

   Plants can break rocks into smaller pieces as their roots grow and push against the rock.
   Plants can cause physical weathering.

Check your answers to questions on Earth and Space Science Test B on pages 59-60.

1. **Most water on Earth is found in**

   B oceans.

   Oceans contain about 97 percent of Earth's water. Only about 3 percent of Earth's water is fresh.

2. **What is the source of energy for the water cycle?**

   C the sun

   Energy from the sun drives the processes of the water cycle, which include evaporation, condensation, and precipitation.

3. **Choose the statement that best describes the water cycle.**

   D During condensation, water changes from a gas to a liquid.

   During condensation, water vapor changes into liquid water. Millions of tiny droplets of water then form clouds.

4. **If a weather forecaster wanted to measure the temperature of air, she would use a(n)**

   A thermometer.

   Thermometers are used to measure temperature. The colored bar of liquid tells what the temperature is.

5. **You would like to live in a climate that is warm year-round. Where should you live?**

   C near the equator

   Tropical regions are near the equator. They receive the most direct sunlight all year and so are warm year-round.

6. **Which type of cloud is shown here?**

   A cirrus

   Cirrus clouds look like feathers.

Short Response

7. **Jose measured the relative humidity outside his classroom several times in one day and recorded his measurements in this chart. When was the air unable to hold more water vapor? Explain your answer.**

   At 2 P.M., the relative humidity was 100%. This means the air couldn't hold more water vapor.

   Air with 100 percent relative humidity is unable to hold more water vapor.

Check your answers for Earth and Space Science Test C on pages 67-68.

1. **Which of the following is an example of a renewable energy resource?**

   D solar energy

   Energy from the sun will not run out for billions of years. So, it is called a renewable energy resource.

2. **Which of the following do we depend on mineral resources for?**

   C electrical wire

   Minerals like copper are used to make electrical wire.

3. **Soil and sand pollute water by**

   B blocking light.

   Particles of sand and soil can block sunlight from reaching plants and animals that live in the water.

4. **A product that shows this symbol on its packaging can be**

   C recycled.

   This is the recycling symbol. Recycling materials helps save resources.

5. **How long does it take for fossil fuels to form?**

   D millions of years

   It takes millions of years for fossil fuels to form.

6. **Where does the energy in fossil fuels come from?**

   D remains of dead plants and animals

   Fossil fuels are formed when the remains of dead plants and animals are buried and pressed together over long periods of time.

Short Response

7. **Why should people use renewable rather than nonrenewable energy resources?**

   It is better to use renewable energy resources because we will not run out of them.

   Once nonrenewable resources are used up, they cannot be replaced. Renewable resources do not run out.

Check your answers for Earth and Space Science Test D on pages 75-76.

1. **All the planets in the solar system orbit the**

   (A) sun.

   The sun is the center of the solar system. All planets revolve around the sun.

2. **The reason that Earth's day is 24 hours long is because**

   (B) it takes Earth 24 hours to make one rotation on its axis.

   As Earth rotates on its axis, half of the globe is facing the sun (day) and half is facing away from the sun (night).

3. **In this picture, the Southern Hemisphere of Earth is experiencing**

   (A) summer.

   The Southern Hemisphere is tilted toward the sun, so it is summer.

4. **The sun rises in the east because**

   (D) Earth rotates from west to east.

   Because Earth rotates from west to east, we see the sun rise in the east and set in the west.

5. **Which planet is closest to the Sun?**

   (C) Mercury

   Mercury is the closest planet to the sun.

6. **Which of the following outer planets is made of solid rock?**

   (C) Pluto

   All the outer planets except Pluto are made of gases. Pluto is made of rock.

7. **Which planet takes the longest time to orbit the sun?**

   (B) Pluto

   Pluto is the farthest planet from the sun. It has the longest orbit.

8. **What is the difference between a solar system and a galaxy?**

   A solar system is made up of a star and all its planets and their moons. A galaxy is made up of billions of stars and their planets.

   A galaxy can have billions of stars, some of which have solar systems.

Properties of Matter

What is matter?

All objects are made of **matter**. Matter has **properties**, or characteristics. Mass, volume, and density are three important properties of matter.

Mass is a measure of the amount of matter in an object. You have more mass than this book. That's because you contain more matter than the book does. You can use a balance to measure mass. An object's mass is measured in milligrams (mg), grams (g), and kilograms (kg).

Volume is a measure of how much space an object takes up. A book has a larger volume than a paper clip. That's because a book takes up more space. You can use a ruler or a graduated cylinder to measure volume. The volume of a rectangular solid is measured in cubic centimeters (cm^3) and cubic meters (m^3). The volume of a liquid is measured in milliliters (mL) and liters (L).

An object's **density** is the amount of mass it has in a given volume. To find the density of an object, divide its mass by its volume. The density of an object is described in grams per cubic centimeter (g/cm^3) or grams per milliliter (g/mL).

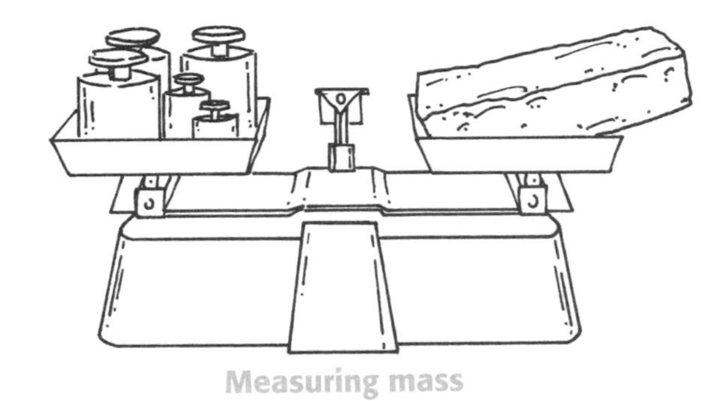

Measuring mass

Measuring volume

Show What You Know

Which block has a higher density? Explain your answer.

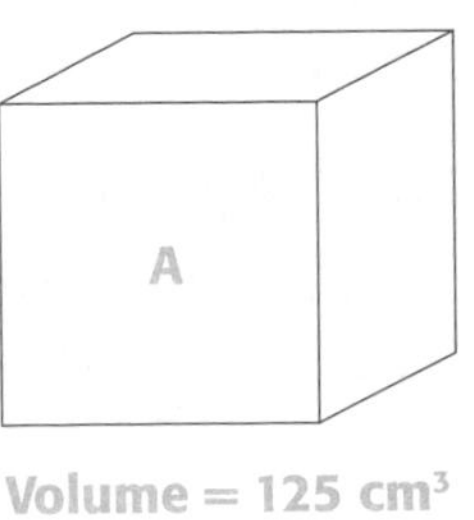

Volume = 125 cm^3
Mass = 20 g

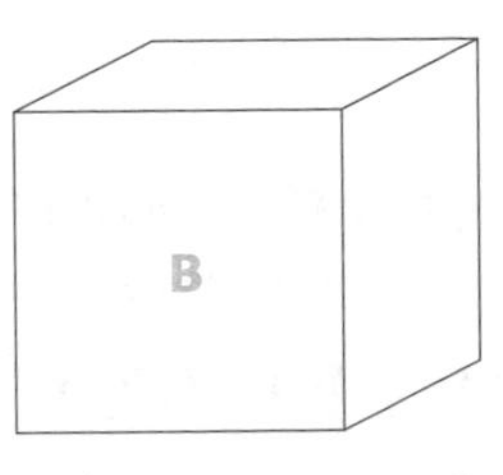

Volume = 125 cm^3
Mass = 28 g

LESSON 54

Properties of Matter

What are some physical and chemical properties of matter?

A **physical property** is a property that can be measured or observed without changing a substance. A **chemical property** describes how a substance can change into another kind of matter.

All substances have their own special set of physical and chemical properties. You can compare the properties of substances to tell them apart. Look at the chart below.

Some Physical and Chemical Properties of Matter

Property of Matter	Type of Change	What It Means
Boiling point	Physical	The temperature at which a substance boils
Melting point	Physical	The temperature at which a substance melts
Conducts heat	Physical	How easily heat moves through a substance
Conducts electricity	Physical	How easily electricity moves through a substance
Is magnetic	Physical	Is attracted to a magnet
Is soluble	Physical	How easily a substance dissolves in another substance
Is buoyant	Physical	If a substance floats or sinks in water
Burns	Chemical	Bursts into flames when heated
Rusts	Chemical	Forms rust when exposed to air and water
Reacts with acids	Chemical	Changes in some way when combined with an acid

Show What You Know

Imagine that you tested the physical and chemical properties of two mystery substances. All of their properties turned out to be the same. What can you conclude about the substances? Explain your answer.

___

___

55

Changes in Matter

How can matter change?

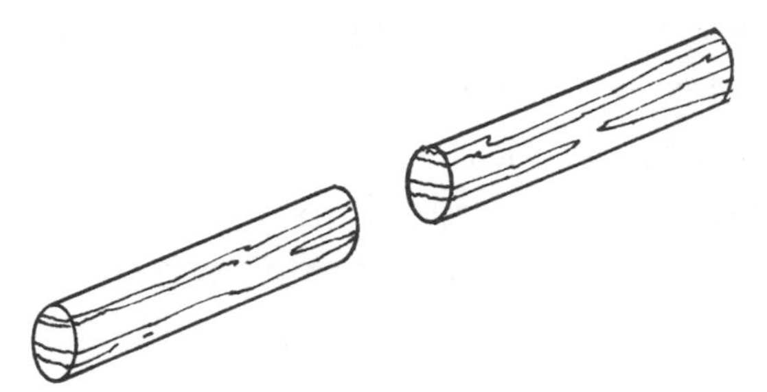

Each piece of the rod has the same properties as the whole rod.

Paper burns to form ash, a new substance with new properties.

A **physical change** is any change to the size, shape, or state of a substance. Let's say a wooden rod is cut in half. Each half is smaller but still has all the properties of wood. Or let's say solid iron is melted into a liquid. The liquid iron still has all the properties of iron, and it can be cooled to form solid iron again.

A **chemical change** produces one or more new substances. Think about what happens when a bike is left outside in the rain for several months. Air and water combine with the iron in the bike to make rust. Rust has different properties than iron. It is a new substance.

Burning is another kind of chemical change. A piece of paper that is burned, for example, turns to ash. Ash has different properties than paper. It is a new substance.

Show What You Know

Identify each change in the chart as physical or chemical. Then complete the chart by adding examples of one physical and one chemical change.

Change	Type of Change
Cutting an apple	
Cooking pancakes	
Burning a log in a fireplace	
	Physical
	Chemical

LESSON 56

Compounds, Mixtures, and Solutions

How are a compound and a mixture different?

Elements are the building blocks of matter. They combine to form all the different kinds of matter there are.

Compounds are substances made of two or more elements. The smallest unit of a compound is a **molecule**. A molecule cannot be changed easily. Sugar is an example of a compound. A molecule of sugar is made of the elements carbon, hydrogen, and oxygen. Like all compounds, sugar has its own set of physical and chemical properties.

Mixtures are made of two or more substances that are not joined together. The substances can be separated easily. Vegetable soup is an example of a mixture. It is made of liquid and different vegetables, such as carrots, onions, and potatoes. Straining the soup separates the vegetables from the liquid. The vegetables can be separated from each other, too. The substances in a mixture keep their own physical and chemical properties.

A **solution** is a kind of mixture. It is made of one substance that is spread out evenly in another substance. Saltwater is a solution. The salt particles are spread out evenly in the water. They are too small to see. But you can taste the salt in the water. And if you heat the water until it evaporates, the salt is left behind.

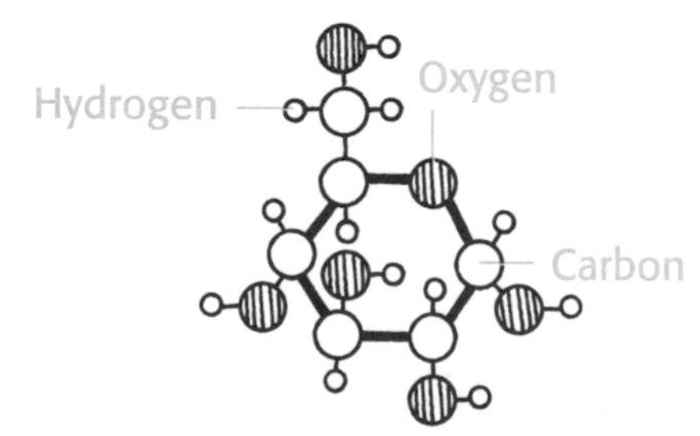

A sugar molecule—A compound

Vegetable soup—A mixture

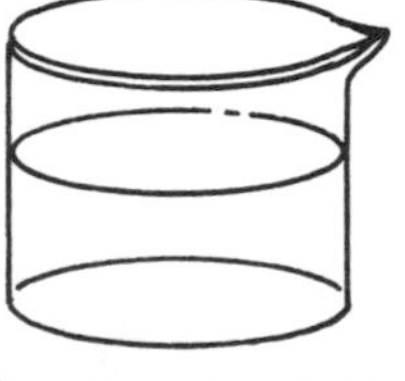

Saltwater—A solution

Show What You Know

Explain how you could separate a mixture of sand and salt. (Hint: Think about the physical properties of each substance. Which properties are different?)

____

____

____

States of Matter

What are states of matter?

Most substances can exist as solids, liquids, or gases. Water is in a liquid state. Ice is water in a solid state. Water also exists in the air as water vapor, a gas. State is a physical property of matter.

A **solid** has a definite shape and a definite volume. The particles in a solid are packed closely together. They do not move around much. So a solid keeps its shape. Wood and rocks are solids.

A **liquid** has a definite volume but not a definite shape. The particles in a liquid are not packed together as closely as in a solid. The particles tumble and flow around each other. That's why liquids take the shape of their containers. Milk, molasses, and shampoos are liquids.

A **gas** does not have a definite shape or volume. The particles in a gas are not packed closely together. They move freely in all directions. That's why gases spread out to fill their containers. Helium, oxygen, and carbon dioxide are gases.

Movement of Particles in Different States of Matter

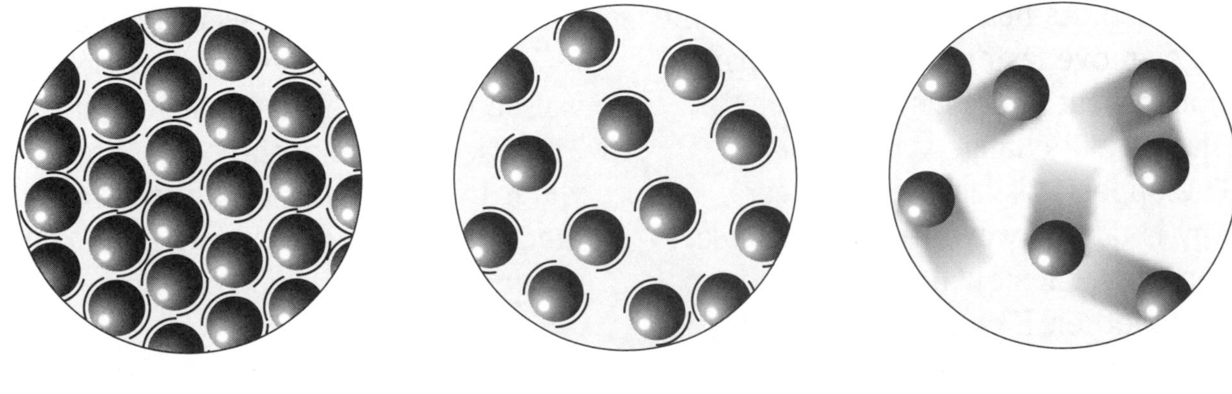

Solid Liquid Gas

Show What You Know

Explain why you can't pour a solid.

____

____

____

LESSON 58

Changing States of Matter

How does matter change state?

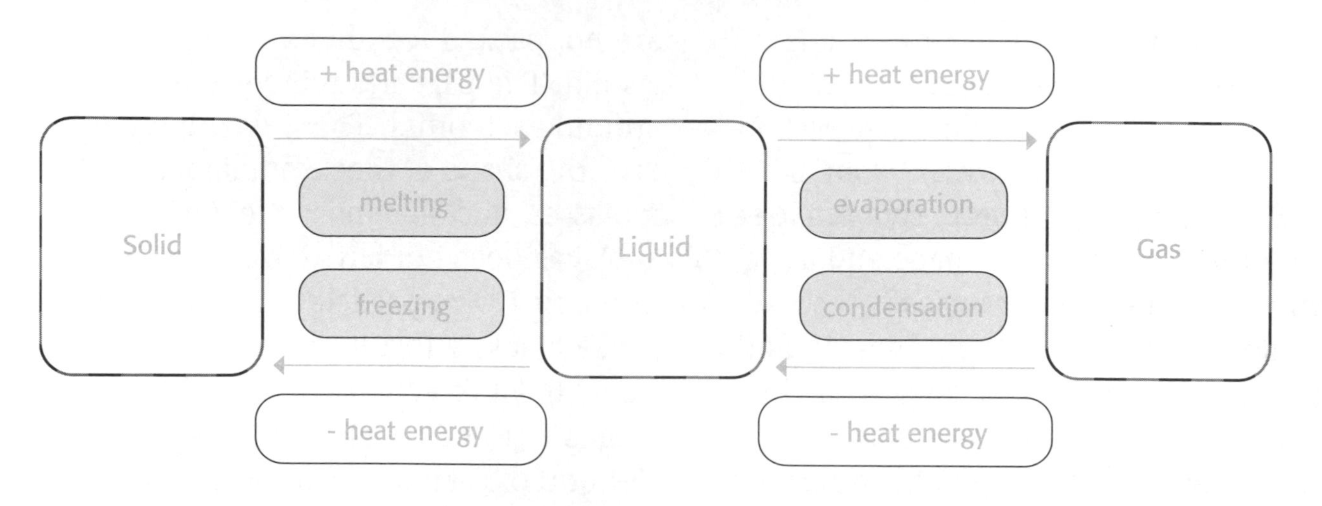

A change of state happens when **heat energy** is added to or taken away from a substance. As heat energy is added, the particles move faster and farther apart. If enough heat energy is added, the substance will change state.

The opposite is also true. If heat energy is taken away from a substance, its particles will slow down and move closer together. If enough heat energy is taken away, the substance will change state. A change of state is a **physical change.** The pictures above show how solids, liquids, and gases change states.

Water is a special substance because it can be found in all three states on Earth. There is solid ice at the poles, liquid water in the oceans, and water vapor in the air. Heat energy from the sun causes water on Earth to change state.

Show What You Know

Describe a time when you saw water change state. What caused the change? Use the term *heat energy* in your answer.

____

____

____

____

Multiple Choice

Fill in the letter to show your answer.

1. **A tool used to measure mass is called a**

   Ⓐ ruler.

   Ⓑ graduated cylinder.

   Ⓒ balance.

   Ⓓ thermometer.

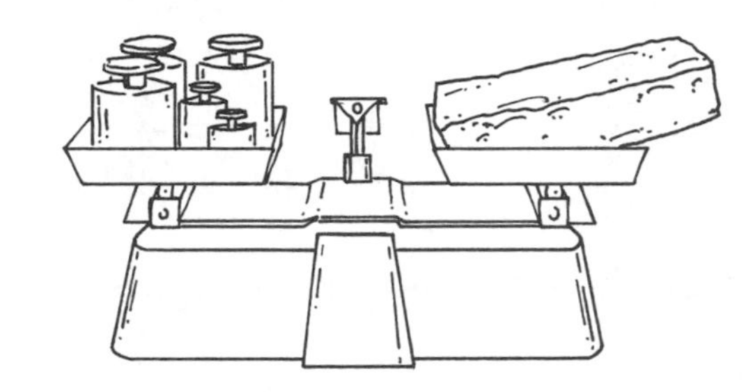

2. **A glass block has a mass of 260 g and a volume of 100 cm³. What is its density?**

   Ⓐ 2.6 g/cm³

   Ⓑ 26 g/cm³

   Ⓒ 100 g/cm³

   Ⓓ 260 g/cm³

3. **Which is *not* a physical property of matter?**

   Ⓐ mass

   Ⓑ volume

   Ⓒ color

   Ⓓ ability to rust

4. **Which is an example of a chemical change?**

   Ⓐ ice melting

   Ⓑ rust forming on an iron bench

   Ⓒ a puddle evaporating

   Ⓓ the slicing of bread

Use the diagram to answer Items 5-7.

Figure 1: Ice

Figure 2: Water

Figure 3: Water vapor

5. **What state is the substance in Figure 2 in?**

   Ⓐ gas

   Ⓑ solid

   Ⓒ liquid

   Ⓓ none of the above

6. **In which figure are the molecules packed together most tightly?**

   Ⓐ Figure 1

   Ⓑ Figure 2

   Ⓒ Figure 3

   Ⓓ The molecules are packed equally tightly in all three figures.

Short Response

7. **Explain what will happen if heat energy is added to the substance in Figure 2.**

   ____

   ____

   ____

   ____

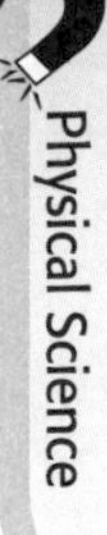

Gravity and Buoyant Force

What are gravity and buoyant force?

A **force** is a push or a pull. Forces can make objects start or stop moving. They can also change an object's direction.

Gravity is the force that attracts objects to each other. The force of gravity between you and Earth is what holds you to Earth's surface. The amount of gravity between two objects depends on two things—how much **mass** each object has and how close together they are. The more mass the objects have, the greater the force of gravity between them. The closer together the objects are, the greater the force of gravity between them.

Mass is the amount of matter in an object. **Weight** is a measure of the pull of gravity on an object. Your weight is a measure of the pull of Earth's gravity on you. An astronaut has the same mass on the moon as she does on Earth. But her weight is only one-sixth what it is on Earth. That's because Earth has about six times more mass than the moon. So the force of gravity on Earth is about six times stronger than the force of gravity on the moon.

Buoyant force is an upward force that acts on objects in water or air. Buoyant force pushes up, while gravity pulls down. Look at the pictures below. Notice that as long as the buoyant force is greater than the pull of gravity, an object, like a boat, will float.

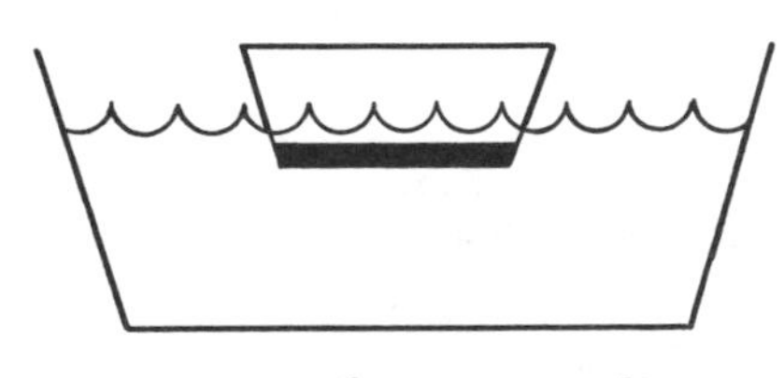

Buoyant force > gravity

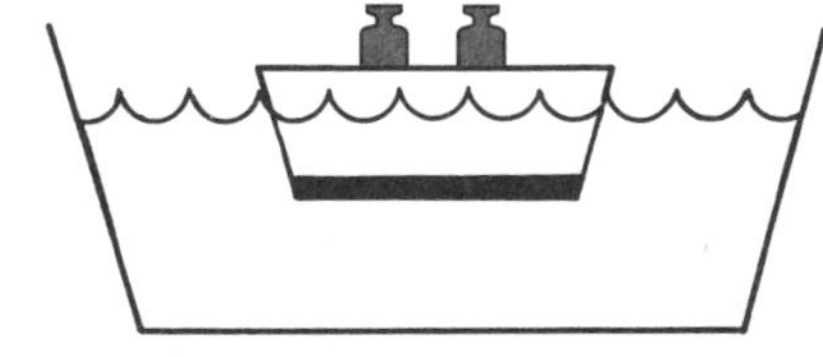

Buoyant force > gravity

Gravity > buoyant force

KEY: > MEANS GREATER THAN

Show What You Know

Explain how a change in the mass of two objects affects the force of gravity between them.

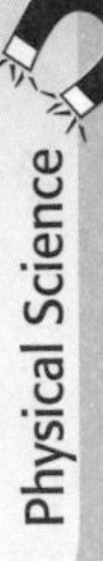

LESSON 60

Magnetism and Friction

What are magnetic force and friction?

A **magnet** is an object that pulls objects made from iron toward it. This pull is called magnetism, or **magnetic force.** You can observe magnetic force when you put two magnets together. Each magnet has a north and a south pole. Opposite, or "unlike," poles pull on each other. The same, or "like," poles push away from each other.

Distance affects magnetic force. If you bring a magnet near an iron nail, the magnet will pull the nail toward it. But the strength of the force goes down the farther away the nail is from the magnet.

Friction is a force that works against motion. For example, when you are riding your bike, you are moving forward. But friction between your tires and the pavement is working against your forward motion. If you stop pedaling, eventually you will stop moving.

Different kinds of objects produce different amounts of friction. The rougher and heavier the objects are, the more friction there is between them. The lighter and smoother the objects are, the less friction there is between them.

Braking on dry pavement

Braking on an icy patch

Show What You Know

Explain why someone would put up this sign after mopping the floors. Use the term *friction* in your answer.

____

____

____

Balanced and Unbalanced Forces

What causes objects to move?

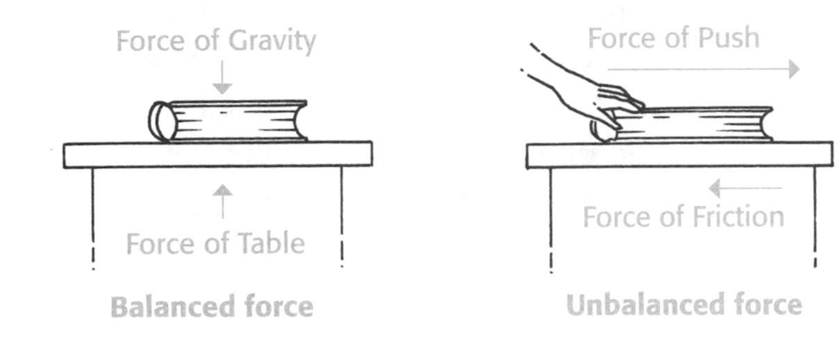

Forces either push or pull objects. If the forces acting on an object are **balanced,** the object does not start or stop moving. Consider a book sitting on a desk. The force of gravity pulls down on the book. But the desk pushes up on the book with equal force. The two forces are balanced. So, the book does not move.

Now imagine pushing one side of the book. The forces acting on the book are now **unbalanced.** The force of the push is working in one direction and the force of friction is working in the other direction. The force of the push is greater than the force of friction, so the book moves in the direction of the push.

In a tug-of-war, two teams pull on opposite ends of a rope. If each team pulls with the same force, the forces are balanced and neither team moves. But if one team pulls harder, the forces become unbalanced.

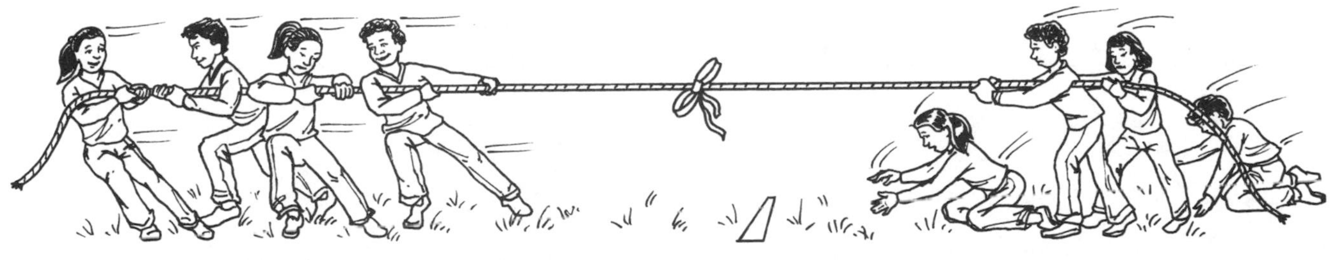
The winning team pulls with a greater force.

Show What You Know

How can you tell if the forces acting on an object are balanced or unbalanced?

____

____

LESSON 62

Motion

What is motion?

Motion is a change in the position of an object. A baseball flying out into left field is in motion. It is changing its position from over the plate to up in the air. The ball's motion can be described by the path it follows as it soars.

Scientists use special terms to describe motion. **Distance** describes how far a moving object travels. It is measured in units such as meters (m) and kilometers (km). Distance is related to speed.

Speed is the distance an object moves in a given amount of time. To find an object's speed, divide the distance it travels by the time it takes to move that distance.

Acceleration is speeding up or slowing down. Or it can be a change in direction. Let's use a sled as an example. As a sled moves down a snowy hill, its speed increases. It accelerates. At the bottom of the hill, it makes a left turn. This change in direction is also acceleration.

Finding the Speed of a Car

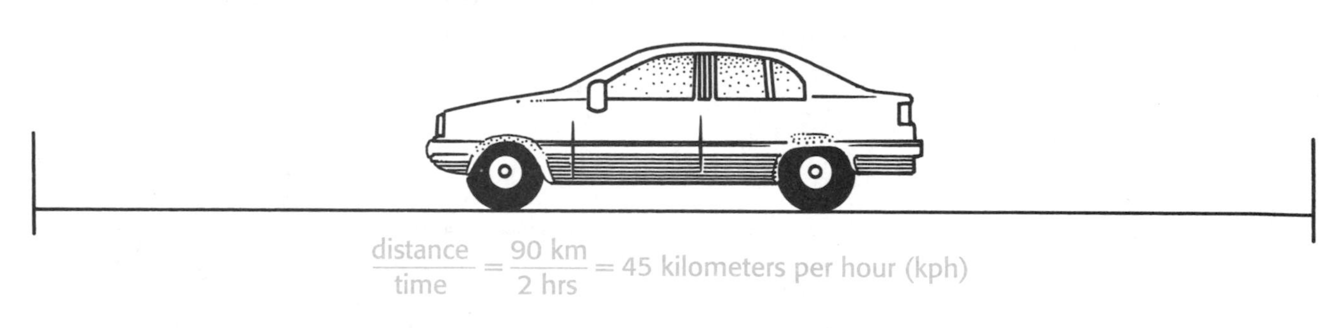

$$\frac{\text{distance}}{\text{time}} = \frac{90 \text{ km}}{2 \text{ hrs}} = 45 \text{ kilometers per hour (kph)}$$

A car traveled 90 kilometers in 2 hours. You can divide to find its average speed.

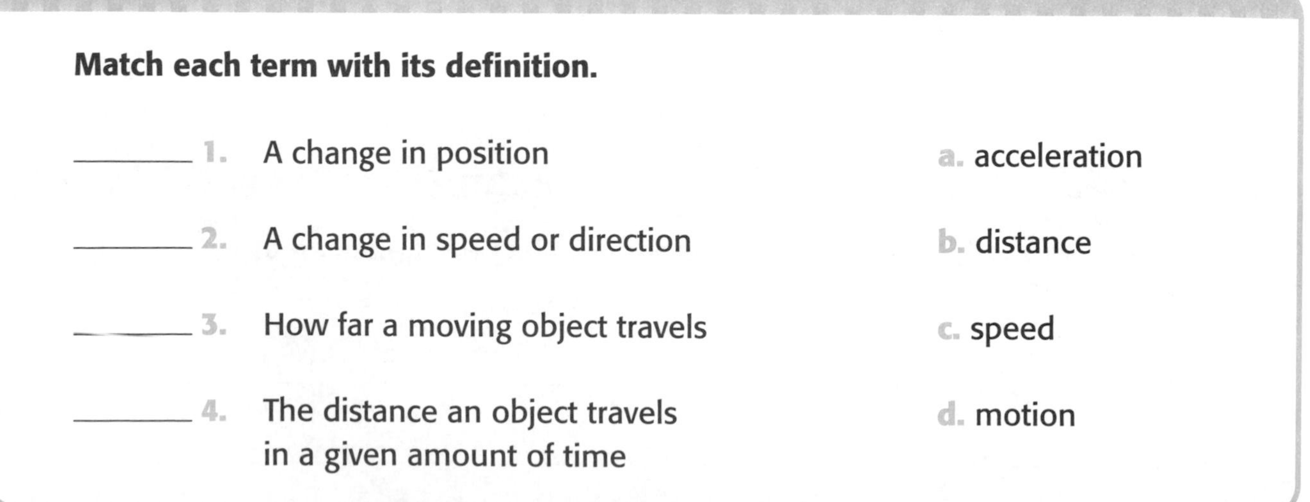

Show What You Know

Match each term with its definition.

_______ 1. A change in position — a. acceleration

_______ 2. A change in speed or direction — b. distance

_______ 3. How far a moving object travels — c. speed

_______ 4. The distance an object travels in a given amount of time — d. motion

Forces and Motion

What forces act on falling objects?

Hold a ball in the air and then let it go. What happens? It falls toward Earth's surface. The force that pulls objects to Earth is gravity. Gravity is a force that can cause motion.

You might think that bigger objects fall faster than smaller objects. Or that a heavier object will fall faster than a lighter object. But if you dropped a marble and a bowling ball from the same height, they would both hit the ground at the same time.

Now imagine dropping two sheets of paper from the same height. One sheet is flat and the other is crumpled. The crumpled sheet would land first. Why? Because air pushes up on falling objects. This upward force is called **air resistance.** A flat sheet of paper faces greater air resistance. It has to push aside more air molecules than a crumpled sheet of paper does.

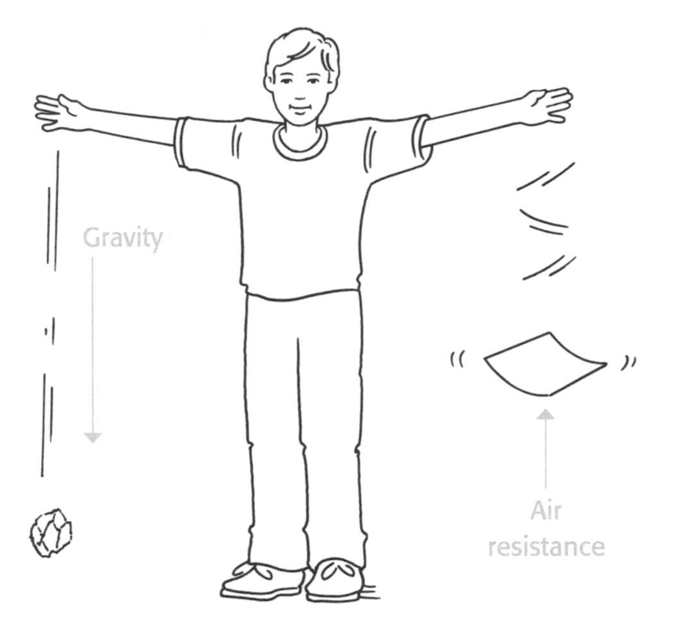

A crumpled sheet of paper falls faster than a flat sheet of paper.

Show What You Know

Draw a picture and use labels to explain how a parachute helps a skydiver land safely on the ground. Use the terms *gravity* and *air resistance* in your work.

Physical Science

LESSON
64

Force, Motion, and Mass

How are force, motion, and mass related?

More force acting on the same mass = more change in motion.

Less force acting on the same mass = less change in motion.

A force acting on an object can change the motion of the object. How much it changes the motion of the object depends on two things—the size of the force and the mass of the object.

A big force changes an object's motion more than a small force. Imagine kicking a soccer ball. A hard kick changes the ball's motion more than a soft kick. Increasing the force makes the ball go farther.

Now imagine the force needed to move two objects with different masses. Think of pushing an empty shopping cart. Your push puts the cart in motion and keeps it in motion. But how does the amount of force you need to use change as the shopping cart fills with heavy groceries? You must use more force to keep the cart in motion.

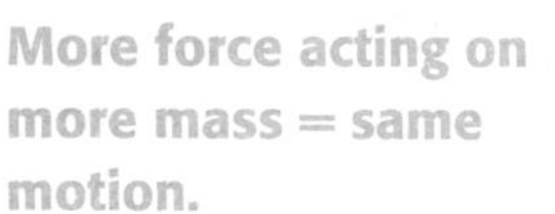

More force acting on more mass = same motion.

Less force acting on less mass = same motion.

Show What You Know

Explain why it is harder to lift a full trash bag than one that is only half-full. Use the terms *force* and *mass* in your answer.

____

____

____

____

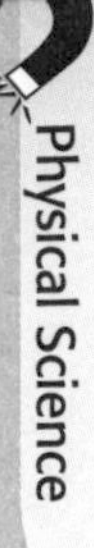

Work and Simple Machines

How do simple machines help us?

Work happens when a force moves an object through a distance. You do work when you push a wheelbarrow across a yard or put a book on a high shelf.

Simple machines make work easier. Most let you use less force to move an object. But in exchange, you have to apply the force for a longer distance.

An **inclined plane** is a ramp, or slanting surface. It is longer than it is high. It takes less force for movers to push heavy boxes up a ramp than it does to lift them into a truck. But in exchange for using less force, they have to push the boxes a greater distance.

A **wedge** is an inclined plane that moves. It is usually made of two inclined planes placed back-to-back. An axe is a wedge used to split wood. It takes less force to split the wood using a wedge. But you must move the wedge through a greater distance.

A **screw** is an inclined plane wrapped around a rod. It takes less force to turn a screw than to pound a nail. But you have to apply the force over a greater distance.

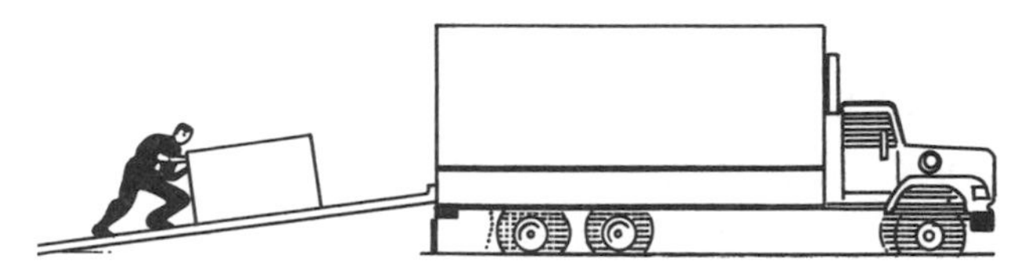

A ramp, or inclined plane

A wedge

A screw

Show What You Know

What kind of simple machine is a spiral staircase most like? Explain how it makes the work of moving to a second floor easier.

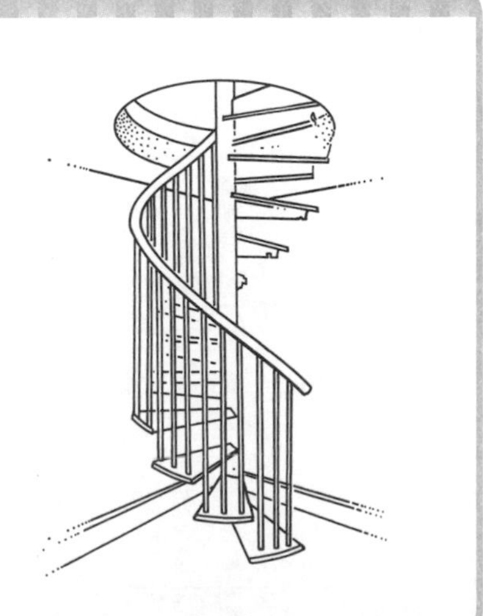

____

____

____

____

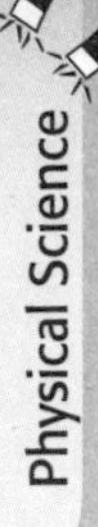

LESSON 66

More Simple Machines

What are some other simple machines?

A **lever** is a bar that turns around a point called a **fulcrum.** There are many kinds of levers, including scissors, wheelbarrows, and seesaws. A lever lets you use less force to lift an object. Look at the picture below. A carpenter is using a crowbar to lift a nail from a piece of wood. She moves the crowbar a greater distance than the nail is moved, but she uses less force to do it.

A **wheel and axle** is a simple machine made of two wheels. A large wheel moves around a small wheel, or axle. A screwdriver is a wheel and axle. Its handle is the wheel, and its tip is the axle. As you twist the handle, the wheel moves a greater distance than the axle. But the force applied to the screw at the axle is greater than the force you had to apply to the handle.

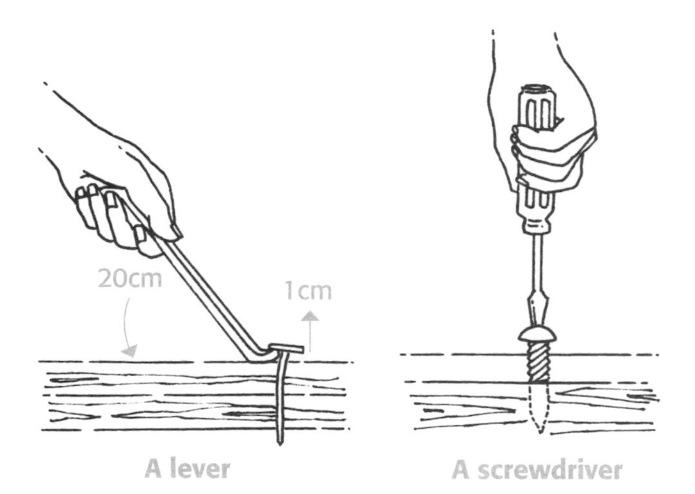

A lever A screwdriver

A **pulley** is a wheel with a rope around it. Pulleys are used to lift heavy objects. A **fixed pulley** simply changes the direction of a force. It allows you to lift a heavy object attached to one end of a rope by pulling down on the other end. A **movable pulley** lets you use less force to lift a heavy object. But you must pull the rope farther than the object is lifted.

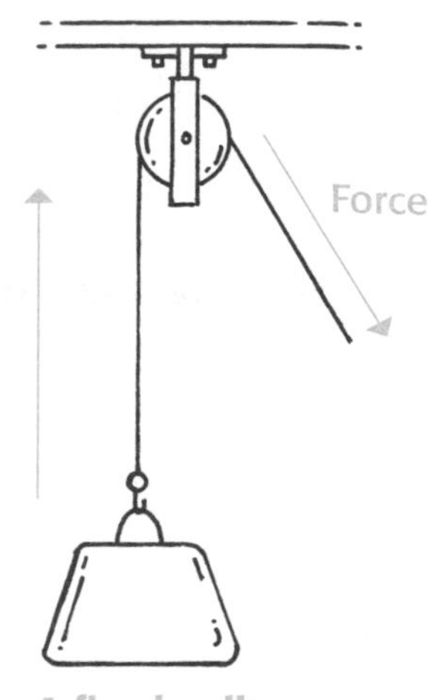

A fixed pulley

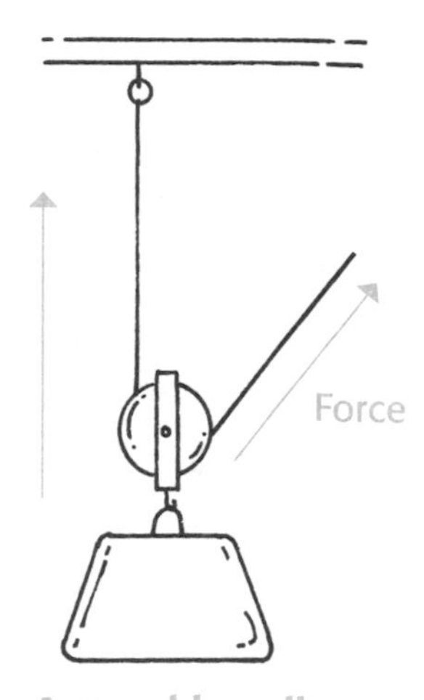

A movable pulley

Show What You Know

Describe a job you could use a fixed pulley to help you do.

____

____

A Multiple Choice

Fill in the letter to show your answer.

1. **Which of the following is *not* an example of a force?**

   Ⓐ gravity

   Ⓑ friction

   Ⓒ magnetic force

   Ⓓ motion

2. **Which force always acts in the opposite direction from gravity?**

   Ⓐ weight

   Ⓑ buoyant force

   Ⓒ magnetic force

   Ⓓ friction

3. **Which of these examples shows an unbalanced force?**

   Ⓐ a rock sitting at the top of a hill

   Ⓑ a rock rolling down a hill

   Ⓒ a dish sitting on a table

   Ⓓ a rug lying on the floor

4. **What is speed?**

   Ⓐ the distance an object travels

   Ⓑ the acceleration of an object

   Ⓒ the position of an object

   Ⓓ the distance an object travels in a certain time

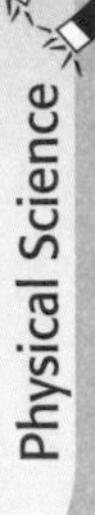

Use the diagram to answer Items 5-6.

5. **What simple machine is being used in the picture?**

   Ⓐ pulley

   Ⓑ wheel and axle

   Ⓒ lever

   Ⓓ inclined plane

6. **How does the simple machine make work easier?**

   Ⓐ It reduce the distance you have to move the lid.

   Ⓑ It reduces the force you need to lift the lid.

   Ⓒ Neither of the above.

   Ⓓ Both of the above.

Short Response

7. **Two magnets are on a table. The poles of one magnet are marked "north" and "south." The other magnet is not marked. How could you find out which pole of the unmarked magnet is the north pole?**

___

___

___

Heat Energy

What is heat energy?

The particles that make up matter are always moving. The energy of moving particles is called **heat energy.**

Like other kinds of energy, heat energy can do work and cause change. For example, it can change the temperature of a substance. When heat energy is added to a substance, its particles start to move faster. **Temperature** is a measure of how fast the particles in matter move. Temperature goes up as the particles move faster. Two common units of temperature are degrees Fahrenheit (°F) and degrees Celsius (°C).

Heat energy can also make a substance change from one state of matter to another. The loss of heat energy is what makes liquid water turn to ice. The addition of heat energy makes a pot of water on the stove start boiling.

Where does heat energy come from? On Earth, it comes mostly from the sun. Heat energy is also produced by other forms of energy. For example, a light bulb gives off heat. You also feel heat when you rub your hands together.

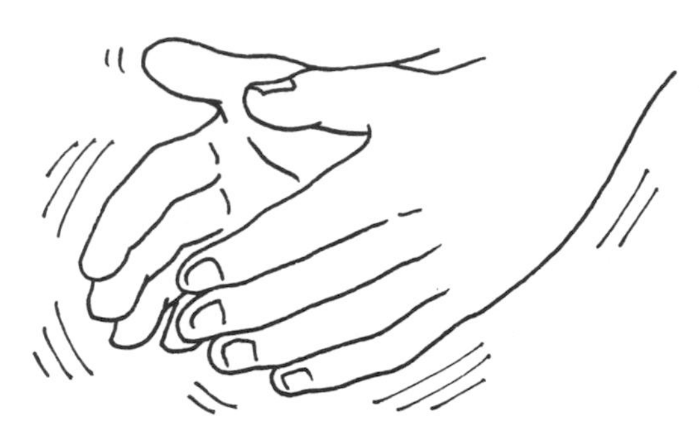

Your hands feel warm when you rub them together.

Show What You Know

A cup of hot chocolate is left sitting on a table. It becomes cool.

1. Explain what happened to the amount of heat energy in the hot chocolate.

____

____

2. Are the particles in the liquid moving faster when the liquid is hot or when it is cool? Explain.

____

____

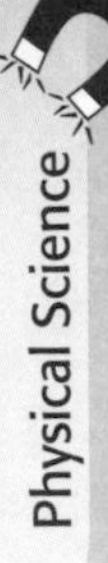

LESSON
68

How Heat Energy Moves

How does heat move from one place to another?

Heat energy always moves from a warmer area or object to a cooler area or object. This kind of movement happens in three ways.

In **conduction,** heat energy moves between two objects that are touching. For example, conduction happens when a cool metal spoon is placed in hot water. Heat energy moves from the water to the spoon. Materials that conduct heat well are called **conductors.** Metals are good conductors. Materials that do not conduct heat well are called **insulators.** Wood and plastic are good insulators.

In **convection,** heat energy moves in currents through a liquid or gas. Let's say you put a pot of water on the stove and turned on the burner. Conduction causes the water at the bottom of the pot to heat up. As the water heats up, its molecules start moving faster. This makes the water lighter, and it rises in the pot. As a current of warm water moves up, cooler water at the top moves to the bottom. Heat energy moves from warmer to cooler water until all of the water becomes hot.

In conduction and convection, heat energy moves through matter. The movement of heat energy through space is called **radiation.** Sit near a campfire, and you can feel heat radiate from the fire. Heat energy from the sun radiates to Earth.

Conduction

Convection

Radiation

Show What You Know

Would you want a pair of mittens made from a conductor or from an insulator? Explain your answer.

____

____

____

LESSON
69

Electrical Energy

What is electrical energy?

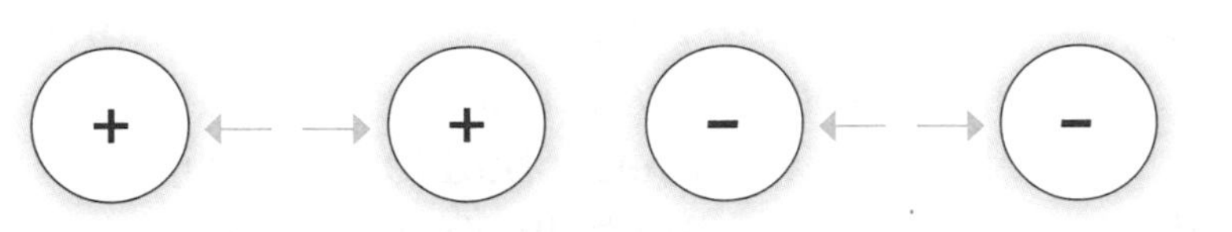

Like charges push each other away.

Opposite charges pull toward each other.

An **atom** is the smallest particle of a substance. Atoms contain electrons and protons.

An **electron** is one kind of charged particle in an atom. **Protons** are also charged particles.

Charged particles produce **electrical energy.** A particle can have a negative charge or a positive charge. Electrons have a negative (–) charge. Protons have a positive (+) charge.

Charges that are alike repel, or push away from, each other. So electrons push each other away. So do protons. But charges that are different attract, or pull toward, each other. Electrons and protons attract each other.

Unlike protons, electrons are free to move from one atom to another. Moving electrons produce electricity, or electrical energy.

Some substances, like metals, conduct electricity. That is why they are used to make electrical wire. A wire might be covered by plastic because plastic does not conduct electricity easily. Plastic is an insulator.

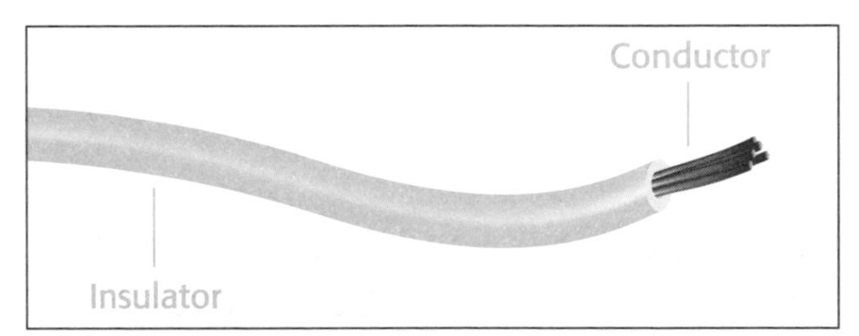

Electrical wire with insulation

Show What You Know

A television screen attracts dust. Do the screen and the dust have the same charge or different charges? Explain your answer.

____

____

____

LESSON 70

Static Electricity

What is static charge?

Most objects have an equal number of positive and negative charges. But electrons can move from one object to another. Objects that lose electrons end up with a positive charge. Objects that gain electrons end up with a negative charge. This buildup of either positive or negative charges on objects is called **static electricity.**

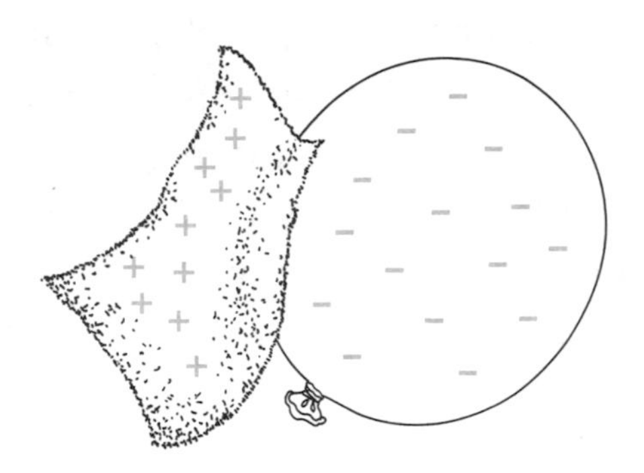

How two objects get a static charge

A piece of wool and a balloon can be used to demonstrate static electricity. The charges on a piece of wool and a balloon are equal until the balloon is rubbed against the wool. Then, electrons move from the wool to the balloon. Now the wool has a buildup of positive charges, and the balloon has a buildup of negative charges. Both the wool and the balloon have a **static charge.**

Two objects with the same static charge repel each other. Two objects with opposite static charges attract each other. Think about clothes tumbling in a dryer. As they tumble, electrons move from one piece of clothing to another. Clothing with different charges stick to each other. That's why you sometimes find your socks stuck to your sweaters.

In time, objects lose their static charges. The loss can be slow or quick. Lightning is an example of a quick loss. During a storm, moving air causes the buildup of charges in clouds. Negative charges in the bottom of a cloud pull toward positive charges on the ground. As electrons move between the cloud and the ground, you see a bolt of lightning.

Show What You Know

You walk across a carpet in your socks and touch a doorknob. A spark flies between your hand and the doorknob. Ouch! What must have happened to your body as you walked across the carpet?

____

____

____

Electric Current

How does electricity flow?

Static electricity is a kind of electrical energy, but it will not play a CD or run a refrigerator. Electric appliances need an **electric current,** or constant flow of electrons. A battery produces enough electric current to run some things. Other things need more current. Power plants produce large amounts of electric current.

To be useful, an electric current must flow through a path. A path for an electric current is a **circuit.** In a circuit, electric current flows through a conductor like a metal wire.

There are two different kinds of circuits. In a **series circuit,** there is only one path the current can flow through. If this path is broken, no current will flow. In a **parallel circuit,** there is more than one path a current can flow through. If one part of the path is broken, the current flows through another path.

You can use a **switch** to turn current on and off. Think of what happens when you turn on a light switch. The light switch completes the circuit, and the bulb lights. Turning the switch off breaks the circuit, and the light goes off.

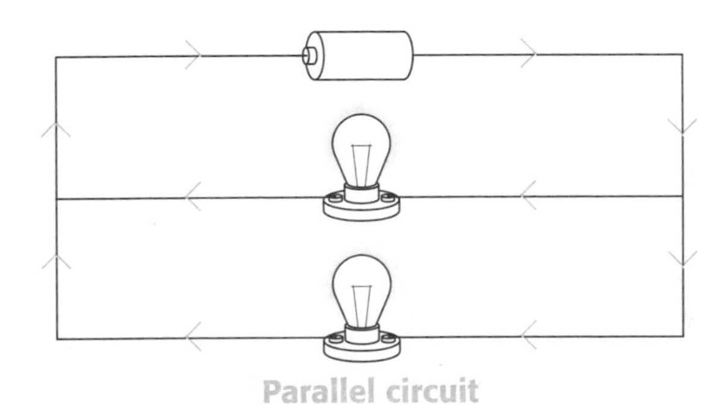

Parallel circuit

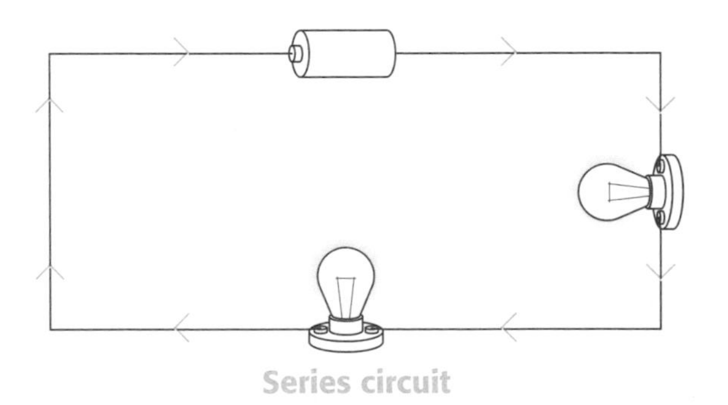

Series circuit

Show What You Know

What kind of circuit is shown in the diagram? How do you know?

____

____

____

____

____

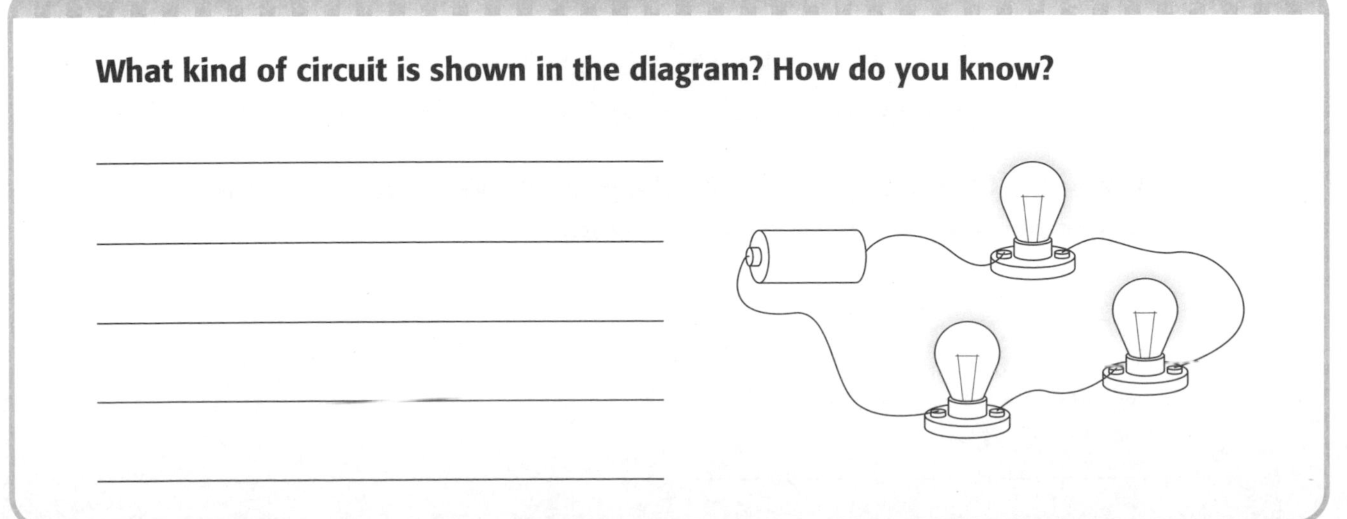

LESSON 72

Magnetism

What is a magnetic field?

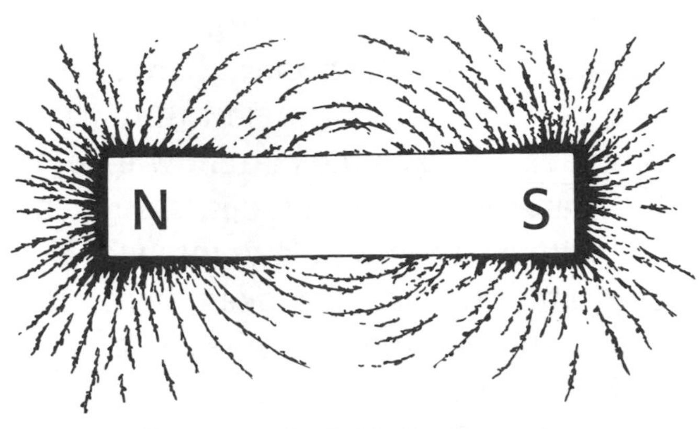

The magnetic field of a magnet

A magnet produces magnetic forces, or **magnetism.** These invisible forces are strongest at a magnet's north and south poles. But they also come from other parts of the magnet, too. Together, they form a **magnetic field.**

You can use iron filings to see a magnetic field. If you put a piece of glass over a magnet and sprinkle iron filings on the glass, the filings will line up along the **lines of force.**

Let's say you put two pairs of magnets under the glass. In one pair, unlike poles face each other. In the second pair, like poles face each other. Just as unlike charges attract each other, so do unlike poles of a magnet. Like poles repel each other. The iron filings show the lines of force between the magnets.

Earth acts as a giant magnet and has a magnetic field. A **compass** is a tool that uses Earth's magnetic field to find direction. Compasses work because the needle of a compass is a magnet. The "north pole" of the needle always points toward Earth's magnetic north pole.

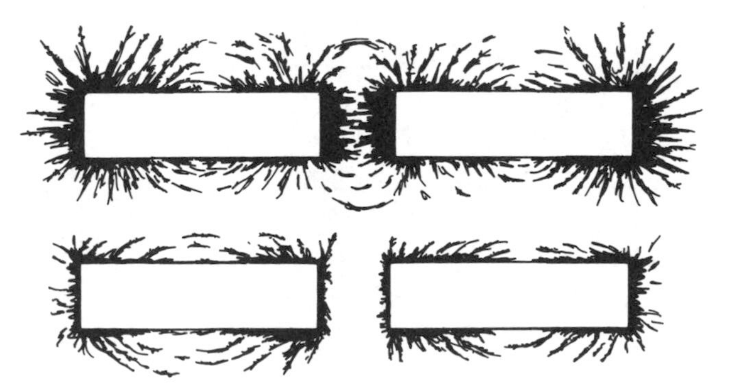

Iron filings follow the lines of force.

Show What You Know

Label the poles of the magnets in the diagram above. How you could tell which poles were north and which were south?

____

____

____

Electromagnetism

How can electricity create magnetism?

Electromagnetism is the relationship between electricity and magnetism. When an electric current passes through a wire, a magnetic field forms around the wire. The wire becomes an **electromagnet.**

Increasing the strength of an electric current through a wire makes a stronger electromagnet. So does coiling the wire around an object like an iron nail. The magnetic fields of each loop in the coil combine to make a strong electromagnet.

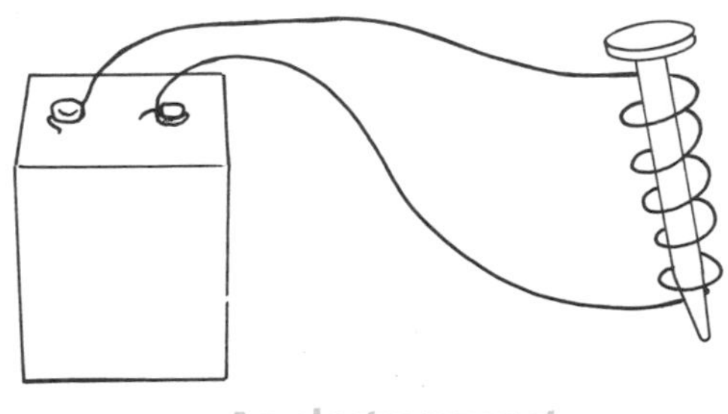

An electromagnet

An ordinary magnet is a **permanent magnet,** which means it is always a magnet. An electromagnet is a **temporary magnet.** That means its magnetism can be turned on and off. Without an electric current, an electromagnet no longer has a magnetic field. It stops being a magnet. Turn the current back on and the wire is a magnet again.

Because they are temporary magnets, electromagnets have many uses. A huge electromagnet, for example, can lift a car in a junkyard or pull steel cans out of a pile of materials. When the current is turned off, the metal drops from the magnet. A permanent magnet could pick up metal, but there would be no way to release the metal from the magnet.

Show What You Know

Which electromagnet is stronger? Explain your answer.

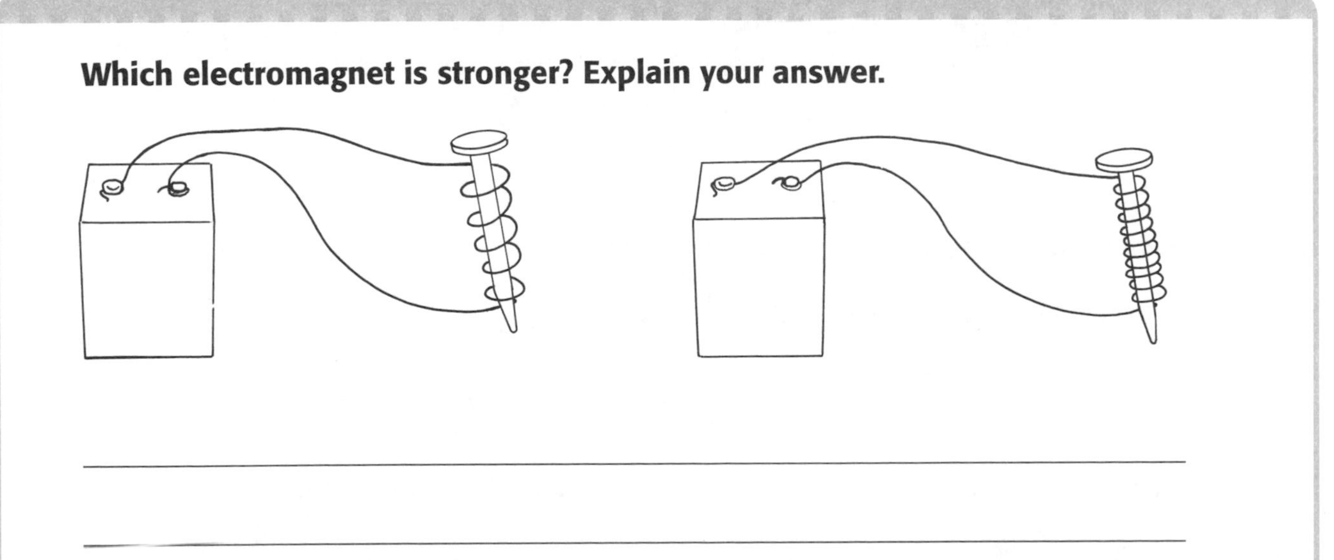

LESSON 74

Uses of Electromagnetism

How can magnetism create electricity?

An electric current can produce a magnetic field. It is also true that a magnetic field can produce an electric current. If a wire is moved through a magnetic field, electrons start to move through the wire. A **generator** produces electricity by moving a loop of wire through a magnetic field. Large generators in power plants produce most of the electricity you use.

An electric **motor** uses electric current to make something move. A motor contains an electromagnet and a permanent magnet. The like poles of the two magnets repel each other. The unlike poles of the magnets attract each other. As a result, the electromagnet turns. As the electromagnet turns over and over again, so do the parts of the motor.

Electromagnets are also used to create sound in doorbells, speakers, and earphones.

Machines with Electric Motors

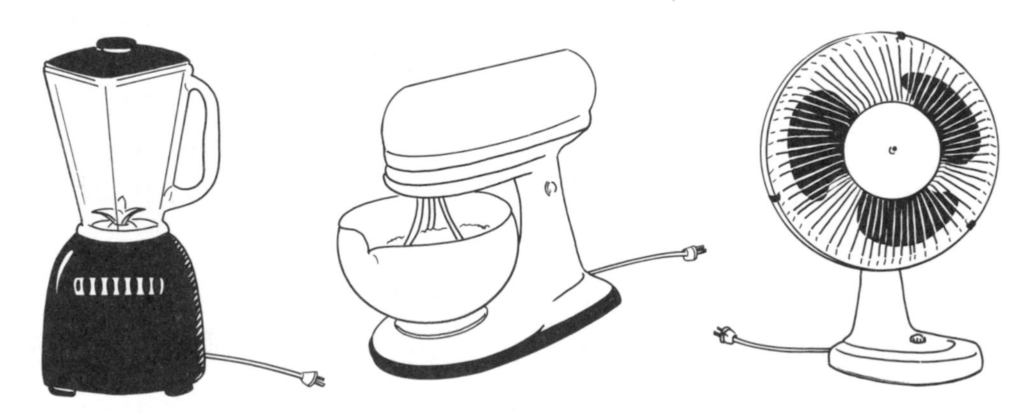

Show What You Know

Use these words to complete the sentences.

electromagnet **generator** **motor**

1. A(n) ____________________ produces electricity from magnetism.
2. A(n) ____________________ uses electric current to cause motion.
3. A(n) ____________________ produces magnetism by running electric current through a wire.

• • PHYSICAL SCIENCE TEST C • •

A Multiple Choice

Fill in the letter to show your answer.

1. **Someone holds a cup of hot chocolate. How does heat move in this example?**

   Ⓐ by conduction

   Ⓑ by convection

   Ⓒ by radiation

   Ⓓ by insulation

2. **The particle that moves in an electric current is the**

   Ⓐ proton.

   Ⓑ neutron.

   Ⓒ electron.

   Ⓓ atom.

3. **In static electricity,**

   Ⓐ protons flow from one place to another.

   Ⓑ charge builds up.

   Ⓒ all items have a negative charge.

   Ⓓ all items have a positive charge.

4. **What kind of circuit is shown?**

   Ⓐ parallel

   Ⓑ series

   Ⓒ static

   Ⓓ open

• • PHYSICAL SCIENCE TEST C • •

5. **Which of the following will *not* make an electromagnet stronger?**

Ⓐ a stronger current in the wire

Ⓑ more loops of wire

Ⓒ wrapping the wire around an object made of iron

Ⓓ a switch

6. **Which of the following changes magnetism into electricity?**

Ⓐ an electric motor

Ⓑ a generator

Ⓒ a doorbell

Ⓓ an electromagnet

Short Response

Use the picture to answer Item 7.

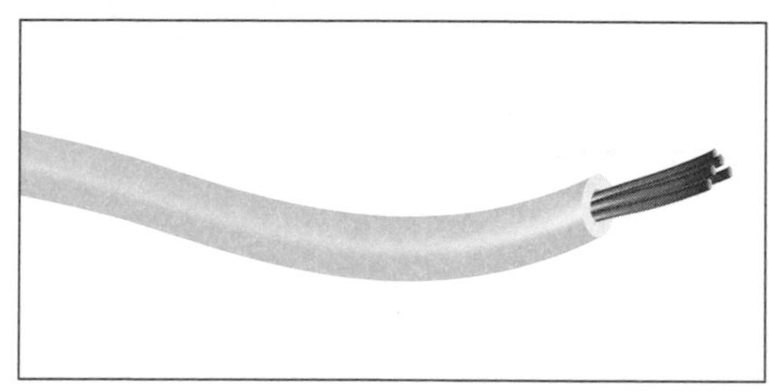

7. **Explain why electrical cords are made of metal wires wrapped in plastic. Use the terms *conductor* and *insulator* in your answer.**

____

____

____

____

____

____

Light Energy

What is light energy?

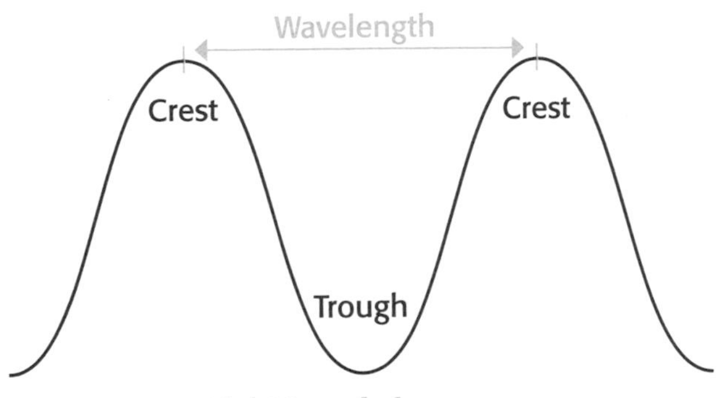

Light travels in waves.

Light is energy that travels in waves through space. Light can come from the sun. Or it can come from a light bulb, candle, or other source. Light waves move away from their source in all directions.

The top of a light wave is the **crest.** The lowest part of a wave is the **trough.** The distance from one crest to the next crest is the **wavelength.**

All light waves travel through air at about the same speed. But some carry more energy than others. That's because some light waves have long wavelengths and others have short ones.

Imagine counting light waves with different wavelengths as they pass by a point. In one second, fewer waves with long wavelengths would pass by the point than waves with short wavelengths. The number of waves that pass by in one second is called a wave's **frequency.** Waves with higher frequencies carry more energy.

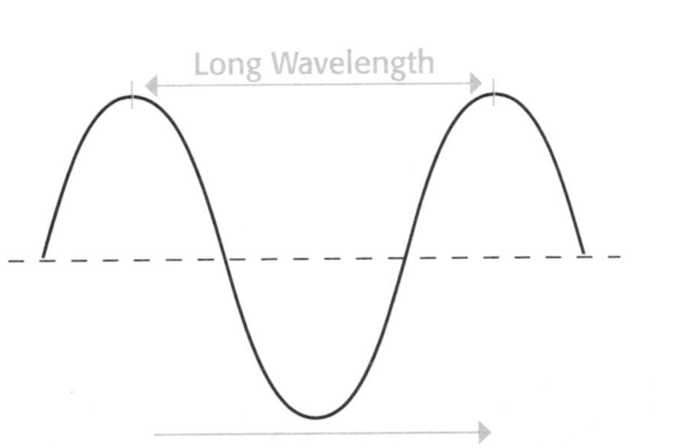

Lower frequency

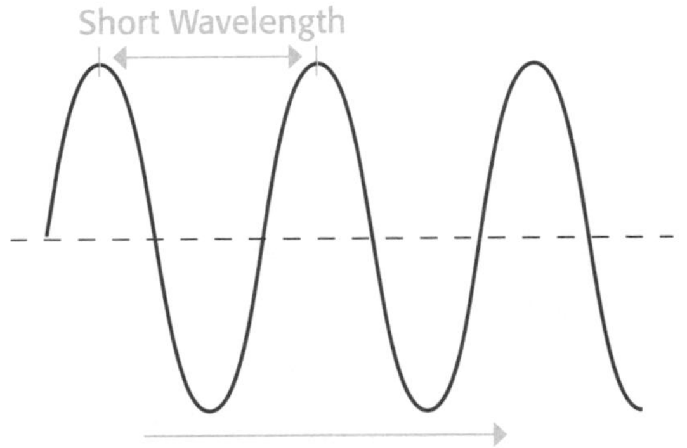

Higher frequency

Show What You Know

Five crests of Wave A pass a point in one second. Nine crests of Wave B pass a point in one second.

1. Which wave has the longer wavelength? ____________________

2. Which wave has the higher frequency? ____________________

3. Which wave carries more energy? ____________________

76

How Light Behaves

What happens when light hits an object?

Light travels in a straight line in all directions from its source. It continues to travel in a straight line until it hits something. When light hits something, different things can happen.

Light can bounce back, or **reflect,** from an object. For example, light reflects from a mirror to form an image of the object in front of the mirror. A light wave reflects at the same angle it strikes.

Light wave

Reflection

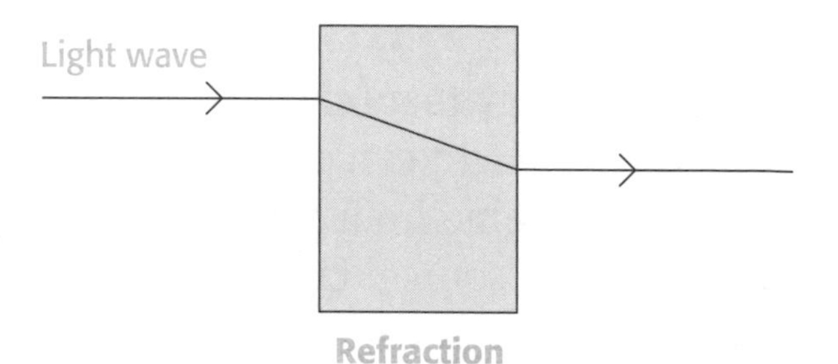

Refraction

Light waves bend, or **refract,** when they leave one material and enter another. The waves refract because they travel at slightly different speeds in different materials. Light, for example, travels faster through air than through water. So light waves refract when they move from air to water, or from water to air.

Light contains all the colors of the rainbow. So why does grass look green and bananas look yellow? When light hits an object, some colors are **absorbed,** or taken in, by the object. Other colors are reflected. The color you see is the color of light that is reflected. Grass is green because it reflects green light and absorbs all other colors. A banana is yellow because it reflects yellow light and absorbs all other colors.

Show What You Know

Look at the picture. What is happening to light waves as they move from the air into the water? How do you know?

____

____

____

____

Sound Energy

What is a sound wave?

Sound is a kind of energy made by objects that vibrate, or move quickly back-and-forth. When objects vibrate, they make molecules in the matter around them vibrate, too. These vibrations form waves that move out in all directions. Sound waves must travel through matter, like air. They cannot travel through empty space.

Think about hitting a drum. The drum cover vibrates. That makes the air molecules around the cover vibrate, too. Those molecules then make the molecules next to them vibrate, and so on. This means that as a sound wave travels, it is the energy of moving air molecules that is passed on. The molecules themselves do not travel from one place to another.

Sound travels most quickly through solid matter. That's because the molecules that make up solid matter are packed together tightly, and so they bump into each other more often. Sound waves travel more slowly through liquid matter, where the molecules don't bump into each other as often. Sound travels slowest through gases.

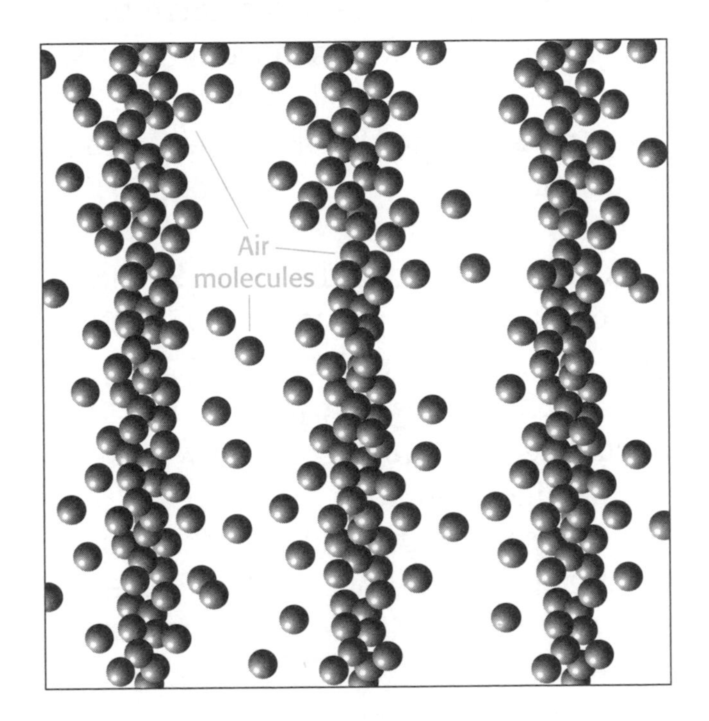

Show What You Know

Imagine two astronauts working on the outside of a space station. One astronaut yells instructions to the other astronaut. Why can't the other astronaut hear her?

____

____

____

LESSON 78

Sound and Hearing

How do you hear sound?

Sound waves travel into your ear and hit your eardrum. The energy of a sound wave makes your eardrum start to vibrate. When the eardrum vibrates, it makes three tiny bones vibrate, too. These vibrations travel to the inner ear, where they make waves in a liquid that is found there. The waves move tiny hairs, which then send signals to the brain. The brain sorts out the signals and tells you what you are hearing and where the sound is coming from.

All sounds are not the same. Sounds can differ in pitch. **Pitch** is how high or how low a sound is. A bird's song has a high pitch. A tiger's growl has a low pitch. Higher-frequency sound waves have a higher pitch.

Some sounds are loud and others are quiet. Larger vibrations produce louder sounds. An explosion makes air molecules vibrate a lot. The sound is loud. A whisper does not make air molecules vibrate much. The sound is quiet.

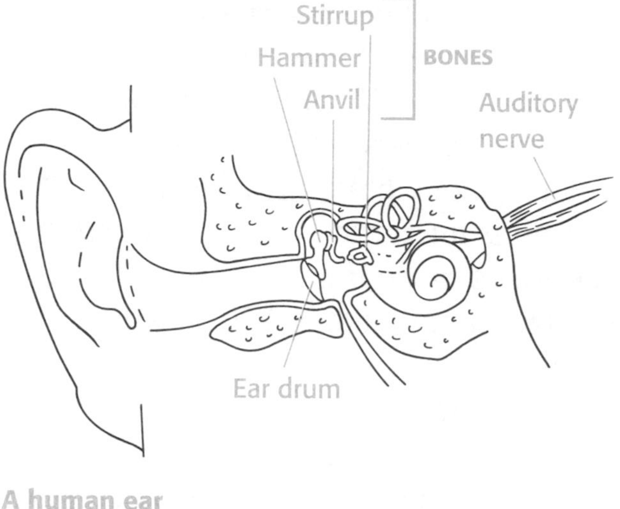

A human ear

Some animals use sound to find food and avoid hitting objects. A dolphin sends out sound waves. If the waves hit something, some of them reflect back to the dolphin as an **echo.** The dolphin uses these waves to know how big the object is, and whether it is food or something dangerous. Bats also use echos to find food and to guide their flying.

Show What You Know

Explain how sound and light are alike and how they are different.

Alike	Different
____________________	____________________
____________________	____________________
____________________	____________________

Mechanical Energy

What is mechanical energy?

Energy related to motion and an object's position is called **mechanical energy.** Some mechanical energy is potential energy. **Potential energy** is stored energy. It comes from an object's position. A rock sitting at the top of a hill is not moving, but it could. It has potential energy. If the rock rolls down the hill, it is moving. A moving object has **kinetic energy.**

A car on a roller coaster has both potential and kinetic energy.

Many objects have both potential and kinetic energy. Let's say a rock starts rolling down a hill. At first, it moves slowly. As it goes down the hill, it moves faster and faster. When the rock was at the top of the hill, its energy was all potential energy because it didn't move. Halfway down the hill, the rock has some kinetic energy. But it also has potential energy because it could move faster. An object's mechanical energy is the sum of its potential and its kinetic energy.

Show What You Know

1. Give an example of how an object's kinetic energy can be changed to potential energy.

___

___

___

2. On the picture of the roller coaster, write the letter **P** on the track where the car has the greatest potential energy. Write the letter **K** where the car has the greatest kinetic energy.

LESSON 80

Energy Changes

How does energy change from one form to another?

It may seem that energy is made and then used up. You turn on a switch and a light bulb comes on. Then you turn off the switch and the light goes off. Or you start a campfire that goes out when the logs are burned up. But energy is not made or used up. Instead, it changes from one form to another. When a light switch is turned on, electrical energy changes into heat and light energy. In a campfire, the logs' chemical energy changes into heat and light energy.

One of the most important energy changes on Earth happens in photosynthesis. **Photosynthesis** is the process plants use to make food. In photosynthesis, light energy from the sun is changed into chemical energy—food. When animals eat plants, they change the chemical energy stored in the food into energy they need to live. Even your car depends on the chemical energy in plants. Gasoline is made from petroleum. Petroleum is made from plants that died millions of years ago.

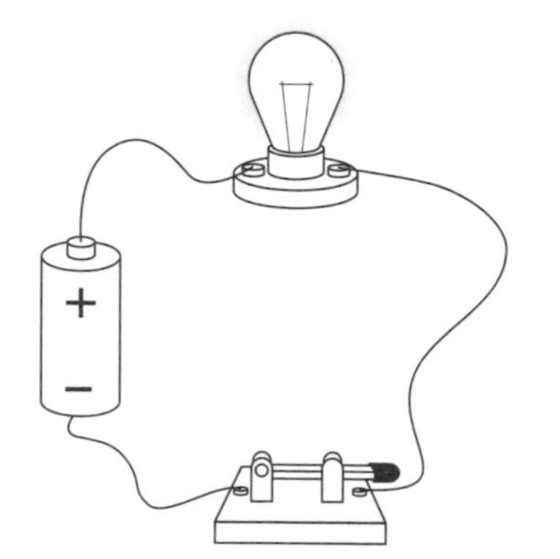

Energy can change from one form to another.

Show What You Know

1. What kind of energy do you think petroleum and gasoline contain? Explain your answer.

____

____

2. Draw a concept map that shows how light energy from the sun becomes the chemical energy in gasoline.

A Multiple Choice

Fill in the letter to show your answer.

1. **A wave with a long wavelength has**
   - Ⓐ a high frequency.
   - Ⓑ a low frequency.
   - Ⓒ more troughs than crests.
   - Ⓓ more crests than troughs.

2. **You can see yourself in a mirror because light**
   - Ⓐ reflects off the mirror.
   - Ⓑ refracts off the mirror.
   - Ⓒ is absorbed by the mirror.
   - Ⓓ bends through the mirror.

3. **Which of the following is true of a sound wave?**
   - Ⓐ It can travel through space.
   - Ⓑ It moves faster through a liquid than a solid.
   - Ⓒ It moves faster through a liquid than a gas.
   - Ⓓ It moves faster through a gas than a liquid.

4. **High-pitched sounds have _________ than low-pitched sounds.**
   - Ⓐ longer wavelengths
   - Ⓑ higher frequencies
   - Ⓒ lower frequencies
   - Ⓓ stronger vibrations

• • PHYSICAL SCIENCE TEST D • •

5. **What kind of energy is stored in petroleum?**

   Ⓐ mechanical

   Ⓑ chemical

   Ⓒ electrical

   Ⓓ heat

6. **The total amount of kinetic and potential energy an object has is its**

   Ⓐ heat energy.

   Ⓑ chemical energy.

   Ⓒ light energy.

   Ⓓ mechanical energy.

Short Response

Use the picture to answer Item 7.

7. **List three energy changes that happen when you use a computer.**

____

____

____

Check your answers to questions for Physical Science Test A on pages 87-88.

Multiple Choice

1. **A tool used to measure mass is called a**

   (C) balance.

   A balance is used to measure mass.

2. **A glass block has a mass of 260 g and a volume of 100 cm³. What is its density?**

   (A) 2.6 g/cm³

   Density is a measure of mass per unit volume.

   $$\frac{260\ g}{100\ cm^3} = 2.6\ g/cm^3$$

3. **Which is *not* a physical property of matter?**

   (D) ability to rust

   The ability to rust is a chemical property of matter.

4. **Which is an example of a chemical change?**

   (B) rust forming on an iron bench

   Rusting is a chemical change. Rust is formed when iron combines with oxygen to form a new kind of matter.

Figure 1: Ice

Figure 2: Water

Figure 3: Water vapor

5. **What state is the substance in Figure 2 in?**

   (C) liquid

   Water is a liquid. Liquids fill the shape of their containers.

6. **In which figure are the molecules packed together most tightly?**

   (A) Figure 1

   The particles of a solid are packed together most tightly of all the states of matter.

Short Response

7. **Explain what will happen if heat energy is added to the substance in Figure 2.**

   Liquid water will evaporate and become a gas called water vapor.

   Adding heat energy causes a change of state. The water in Figure 2 is a liquid. When heat energy is added, the water changes state. It becomes a gas.

• • PHYSICAL SCIENCE TESTS ANSWER GUIDE • •

Check your answers for Physical Science Test B on pages 97-98.

Multiple Choice

1. **Which of the following is *not* an example of a force?**

   D motion

   Motion is not a force. Forces like gravity and buoyancy act upon an object that is in motion.

2. **Which force always acts in the opposite direction from gravity?**

   B buoyant force

   Buoyant force pushes upward. Gravity pulls downward.

3. **Which of these examples shows an unbalanced force?**

   B a rock rolling down a hill

   If the forces acting on an object are unbalanced, the object is moving.

4. **What is speed?**

   D the distance an object travels in a certain time

   Speed is the distance an object moves in a given amount of time.

5. **What simple machine is being used in the picture?**

   C lever

   The screwdriver is being used as a lever to open the can.

6. **How does the simple machine make work easier?**

   B It reduces the force you need to lift the lid.

   A lever makes work easier by reducing the force needed to lift objects. In exchange for using less force, you have to move the lever through a distance greater than the object moves.

Short Response

7. **Two magnets are on a table. The poles of one magnet are marked "north" and "south." The other magnet is not marked. How could you find out which pole of the unmarked magnet is the north pole?**

   I could hold the north end of the marked magnet near one end of the unmarked magnet. If the two ends push each other away, I know that end is the north pole.

   Like poles of a magnet push away from each other. Unlike poles pull on each other.

Check your answers for Physical Science Test C on pages 107–108.

A Multiple Choice

1. **Someone holds a cup of hot chocolate. How does heat move in this example?**

   A by conduction

   The cup and the hands touch each other. Conduction is the movement of heat between objects that are touching. Heat always moves from the warmer object to the cooler object.

2. **The particle that moves in an electric current is the**

   C electron.

   Electric current is the flow of electrons through a path.

3. **In static electricity,**

   B charge builds up.

   In static electricity, a charge builds up and stays on an object.

4. **What kind of circuit is shown?**

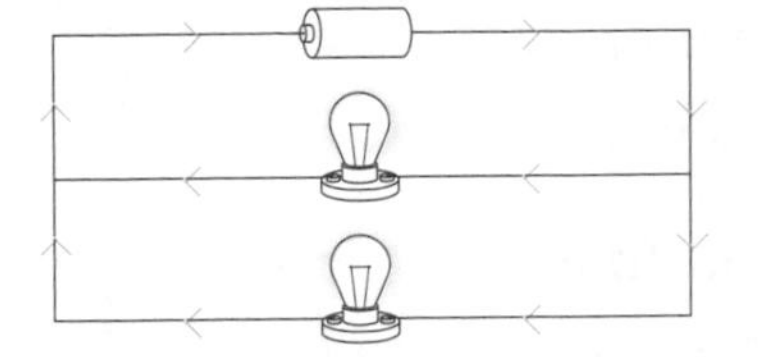

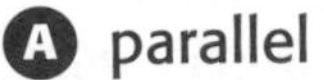

   A parallel

   A parallel circuit provides more than one path for an electric current.

5. **Which of the following will *not* make an electromagnet stronger?**

   D a switch

   A switch completes or breaks a circuit. It can be used to turn an electromagnet on and off, but it does not change its strength.

6. **Which of the following changes magnetism into electricity?**

   B a generator

   In a generator, a wire moves through a magnetic field, producing an electric current in the wire.

Short Response

7. **Explain why electrical cords are made of metal wires wrapped in plastic. Use the terms *conductor* and *insulator* in your answer.**

   Metal is a conductor. Electrical wires are made of metal because electricity moves through them easily. They are covered in plastic, an insulator, because insulators do not let electricity pass through them easily. So the plastic protects people who touch the cord.

   Conductors like metals let electricity move easily. Insulators don't, so insulators like plastic are wrapped around conductors to keep them safe to touch.

Check your answers for Physical Science Test D on pages 115–116.

A Multiple Choice

1. **A wave with a long wavelength has**

   (B) a low frequency.

   If a wave has a long wavelength, not many waves will pass by a point in one second.

2. **You can see yourself in a mirror because light**

   (A) reflects off the mirror.

   Light reflects, or bounces back, after it strikes a mirror.

3. **Which of the following is true of a sound wave?**

   (C) It moves faster through a liquid than a gas.

   Sound travels faster through substances whose molecules are packed tightly together. That means sound travels faster through liquids, such as water, than gases, such as air.

4. **High-pitched sounds have _______ than low-pitched sounds.**

   (B) higher frequencies

   Sound waves with higher frequencies have a higher pitch.

5. **What kind of energy is stored in petroleum?**

   (B) chemical

   Chemical energy is stored in the food made by plants during photosynthesis. Petroleum is made from plants that died millions of years ago.

6. **The total amount of kinetic and potential energy an object has is its**

   (D) mechanical energy.

   Mechanical energy is the total amount of potential and kinetic energy an object has. Potential energy is stored energy that comes from an object's position. Kinetic energy is the energy a moving object has.

Short Response

7. **List three energy changes that happen when you use a computer.**

   The computer runs on electrical energy. Electrical energy changes into sound and light energy to see the screen. The computer gets hot, too. That means electrical energy changes to heat energy.

   Answers could include the change of electrical energy into light on the computer screen, sound, and heat. The kinetic energy of the computer's fan also changes to heat and sound.

LESSON 81

Scientific Inquiry

How do scientists work?

To learn about the natural world, scientists ask questions. Then they do careful investigations to find the answers to those questions. This process is known as **scientific inquiry.**

A **scientific investigation** is the search for an answer to a scientific question. A scientific question is clear and well defined. It has the same meaning for everyone.

Step	What You Do in This Step
1. Ask a good question.	Ask *How, What, When, Where,* or *Why*.
2. Research the topic.	Learn more about the topic you are researching.
3. Make an **hypothesis.**	Suggest an hypothesis, or possible answer to the question.
4. Plan an investigation or experiment.	Decide what steps you will take to find the answer to your question.
5. Conduct the investigation or experiment.	Carry out the steps of your investigation or experiment.
6. Collect and record **data**, or information.	Make observations. Collect and organize data in lists, drawings, or tables.
7. Examine the data and draw conclusions.	Decide what the data mean and if they support the hypothesis.

Show What You Know

What step in the process of scientific inquiry are the students in the picture doing?

____

____

____

____

____

____

LESSON 82

Forming an Hypothesis

What is an hypothesis?

Once you have a question you want to answer, you must form an hypothesis. An **hypothesis** is an idea about what might happen in an experiment. The idea might be based on research you have done. Or it might be based on something you learned from another experiment. You then do an experiment to see if your hypothesis is correct.

Show What You Know

Identify the hypothesis being tested in each of the experiments below.

1. Cups A and B are filled with the same kind of soil. A healthy lima bean is planted in each cup. Both cups are watered equally. Cup A is put by a sunny window. Cup B is put in a dark closet. The plants are checked after one week.

   Hypothesis:

   ____

   ____

2. One tray holds grass growing in soil. One tray holds plain soil with no grass. Both trays are tipped up on one side. Water is poured down over the trays. Runoff is collected at the bottom of the trays.

   Hypothesis:

   ____

   ____

3. One skateboard has a brick on it and one doesn't. Both skateboards are placed on a ramp and let go at the same time. The distance traveled by each skateboard is measured with a meterstick.

   Hypothesis:

   ____

   ____

4. Bowl A and Bowl B are filled with the same amount of water. Bowl A is covered in plastic wrap. Both bowls are left on a counter. The water level is checked in both bowls after one week.

   Hypothesis:

   ____

   ____

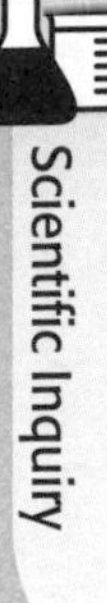

Designing an Experiment

How do you design an experiment?

The purpose of an experiment is to answer a question, or hypothesis. To begin an experiment, you must plan a **procedure,** or list of steps, you will follow. These steps should help you find the answer to your question. Decide what materials you will need and write them in a list. A list makes it easier to collect the things you need. It also tells other people what materials you used in your experiment.

An important part of planning your experiment is controlling the variables. A **variable** is anything that can change in an experiment. You must keep all variables, except the one you are testing, the same. The variable you change is the **independent variable.** When you control variables, you know the results of your experiment are caused by the variable you tested.

Good experiments are done more than once. Each repeated experiment is called a **trial.** Doing many trials helps you be sure that the answers you get are correct.

Question
How does exercise affect heart rate?

Hypothesis
Exercise increases heart rate.

Experiment Procedure

1. Sit quietly for 5 minutes.
2. Measure and record your pulse (heart rate).
3. Do 5 minutes of exercise:
   - walk for 1 minute, or
   - run in place for 2 minutes, or
   - do jumping jacks for 2 minutes.
4. Measure and record your heart rate.
5. Sit quietly again for 5 minutes.
6. Measure and record your heart rate.
7. Repeat steps 1–6 each day for 4 more days.

Show What You Know

Read the procedure described in the flowchart. Identify and list the materials you would need to do this experiment.

____

____

LESSON 84

Variables

What is an independent variable?

An experiment tests only one variable. This variable is called the **independent variable.** Other variables in an experiment are carefully controlled. You must keep these variables the same so that you can test the effect of the independent variable.

Show What You Know

Indentify the independent variable being tested in each of the experiments below.

1. Cups A and B are filled with the same kind of soil. A healthy lima bean is planted in each cup. Both cups are watered equally. Cup A is put by a sunny window. Cup B is put in a dark closet. The plants are checked after one week.

   Independent Variable:

   ____

   ____

2. One tray holds grass growing in soil. One tray holds plain soil with no grass. Both trays are tipped up on one side. Water is poured down over the trays. Runoff is collected at the bottom of the trays.

   Independent Variable:

   ____

   ____

3. One skateboard has a brick on it and one doesn't. Both skateboards are placed on a ramp and let go at the same time. The distance traveled by each skateboard is measured with a meterstick.

   Independent Variable:

   ____

   ____

4. Bowl A and Bowl B are filled with the same amount of water. Bowl A is covered in plastic wrap. Both bowls are left on a counter. The water level is checked in both bowls after one week.

   Independent Variable:

   ____

   ____

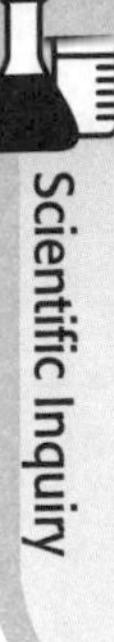

Observations and Data

How are data collected and recorded?

An **observation** is information you use your senses to collect. Your senses include sight, hearing, touch, smell, and taste. The information you collect during an investigation or experiment is your **data.**

Many tools help you make observations. A hand lens, for example, lets you see details too small to see using only your eyes. Rulers and scales let you measure length and mass.

You should record, or write down, all the data from an investigation. There are many ways to record data. You can write observations in lists or sentences. You can draw what you see. Or you may record data in a table. A table shows data in an organized way. Think about how you will use the information to decide the best way to record your data.

To keep data accurate, record observations as you make them. Be sure to include units for your measurements. Keep your data honest by not changing the numbers to make the results come out the way you think they should.

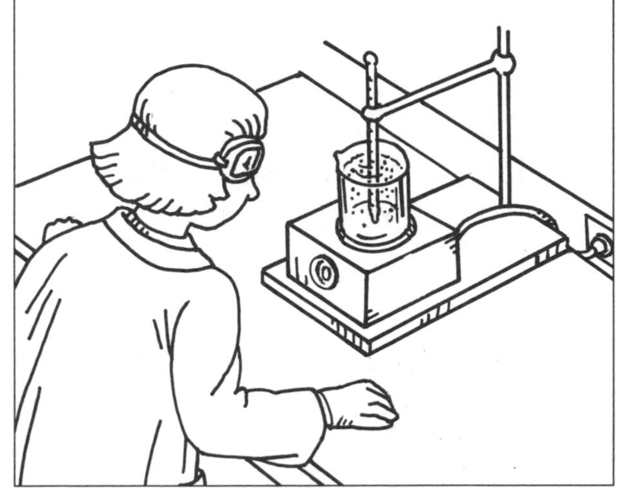

A student observes a beaker of boiling water. Later, she records her observations in a table.

Time	Temp (°C)	Observation
start	22°	water is clear, no bubbles
after 1 minute	45°	a few tiny bubbles forming
after 2 minutes	78°	more bubbles starting to form
after 3 minutes	100°	water boiling

Show What You Know

Use these words to complete the sentences: observation data record

1. When you ____________________ information, you write it down.

2. The information you gather in an experiment is your ____________________.

3. You use your senses to make an ____________________.

LESSON 86

Reading Data Tables

How do you read a data table?

A **data table** organizes the data collected in an experiment. A data table has rows and columns, which are labeled to describe the data they show.

The data table below shows the data collected in an experiment to test the effect of light on the growth of seedlings. You can also use the data in the table to find other information, like how much the seedlings grew over time.

Seedling Growth

		Stem Height (cm)				
Seedling Number	Hours of Light Daily	Day 1	Day 4	Day 8	Day 12	Total Growth
1	0	0.5	0.5	0.5	0.5	0.5 - 0.5 = 0.0
2	3	0.5	0.7	0.7	0.8	0.8 - 0.5 = 0.3
3	6	0.5	0.9	1.2	1.8	
4	9	0.5	1.9	2.4	3.0	
5	12	0.5	2.3	3.1	3.9	

Show What You Know

1. Use the data in the table to determine how much each seedling grew over 12 days. Record your data in the table. The first two have been done for you.

2. How does light affect seedling growth? Use the data in the table to support your answer.

____

____

____

Multiple Choice

Fill in the letter to show your answer.

1. **Look at the experimental setup below. Which of the following is most likely the hypothesis for this experiment?**

   Ⓐ Saltwater has a lower boiling point than freshwater.

   Ⓑ Saltwater has less mass than freshwater.

   Ⓒ Saltwater contains more minerals than freshwater.

   Ⓓ Saltwater does not freeze.

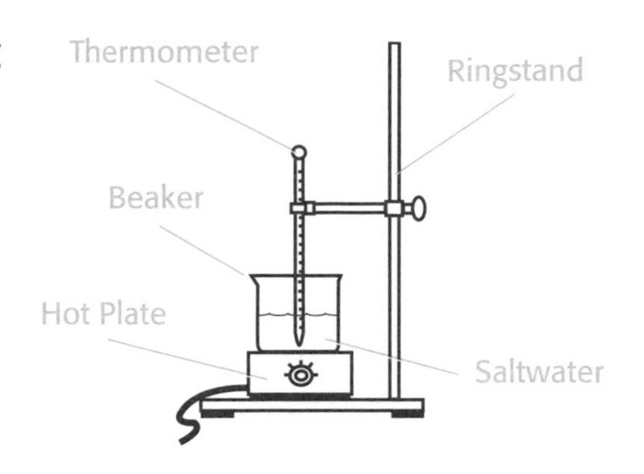

2. **A student plans an investigation about plant growth. The procedure has four trials. The type of plant, type of soil, and amount of water remain the same in each trial. Each group of plants receives different amounts of light. The independent variable in this experiment is the**

   Ⓐ soil type.

   Ⓑ amount of light.

   Ⓒ amount of water.

   Ⓓ type of plant.

3. **Dylan does an experiment. The data he gathers do not support his hypothesis. Dylan should**

   Ⓐ change his data to make them support his hypothesis.

   Ⓑ change his hypothesis to match his data.

   Ⓒ report his results exactly as he observed them.

   Ⓓ repeat the experiment and then change his results.

Short Response

Use the information below to answer Items 4–6.

A student reads about the planets. He makes a data table to show the information he gathered.

Comparing Planet Distances and Orbit Time

Planet	Average Distance from the Sun (km)	Orbit Time (Earth years)
Mercury	57,909,100	0.24
Venus	108,208,600	0.62
Earth	149,598,000	1.00
Mars	227,939,200	1.88
Jupiter	778,298,400	11.86
Saturn	1,427,010,000	29.46
Uranus	2,869,600,000	84.01
Neptune	4,496,700,000	164.80
Pluto	5,913,490,000	248.53

4. **Which planets are farther from the sun than Earth is?**

____

____

5. **Which planet can orbit the sun four times in the amount of time it takes Earth to orbit the sun once?**

____

6. **Compare the distance of the planets from the sun to the time it takes each planet to orbit the sun. What pattern can you see?**

____

____

____

Making a Bar Graph

How do you make a bar graph?

A **bar graph** uses bars of different lengths to compare data. It lets you compare data at a glance. The data table to the right shows the densities of different substances.

Densities

Substance	Density (g/cm^3)
Water (liquid)	1.0
Rubber	1.3
Lead	11.3
Steel	7.8
Aluminum	2.7

Show What You Know

Follow the steps below to make a bar graph of the data in the table. Some steps have been done for you.

Step 1 Think of a title that describes what the graph is showing. Write the title above the graph.

Step 2 Label each axis, or side, of your graph. Here, the horizontal axis shows the names of the different substances. The vertical axis shows density measurements. Label the vertical axis "Density (g/cm^3)."

Step 3 Choose a scale. In this case, the scale goes from 0 to 12, a number slightly larger than the highest number in the data.

Step 4 Draw a bar for each substance. Line up the top of each bar with its correct density on the vertical axis. Label each bar.

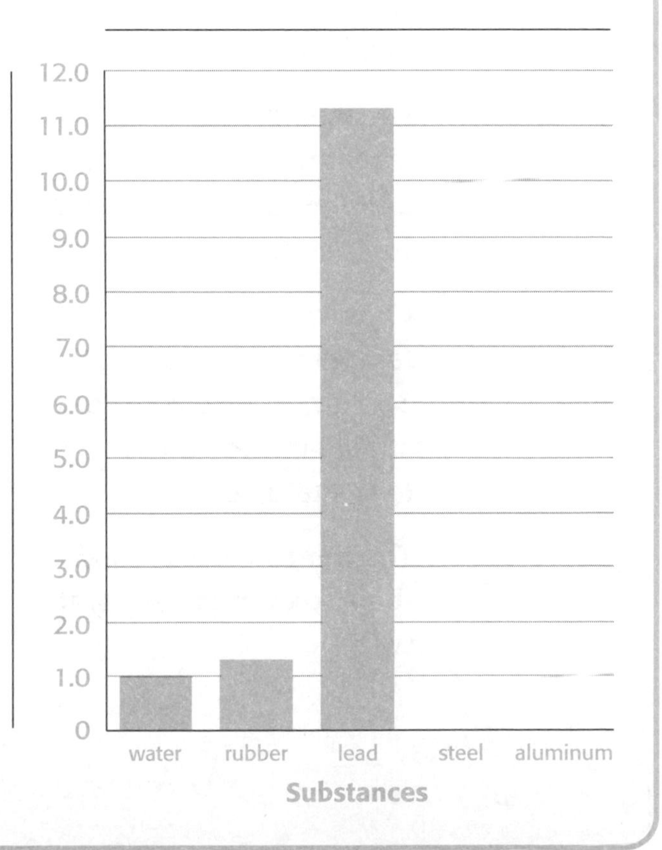

LESSON 88

Making a Line Graph

How do you make a line graph?

A **line graph** is a graph that shows how data change, usually over time. The data table to the right shows the highest temperatures recorded each day over one week.

Daily High Temperatures

Day of Week	Temperature °F
Sunday	37°
Monday	35°
Tuesday	34°
Wednesday	32°
Thursday	30°
Friday	32°
Saturday	35°

Show What You Know

Follow the steps below to make a line graph using the data in the table. Some steps have been done for you.

Step 1 Title the graph. The title should describe what the graph is showing.

Step 2 Label each axis. In this case, label the horizontal axis *Day of Week* and the vertical axis *Temperature* (°F).

Step 3 Choose a scale. The temperatures range from 30°F to 40°F, so your scale should include these temperatures.

Step 4 The horizontal axis shows the abbreviated days of the week.

Step 5 Draw, or plot, a point where the lines for Sunday and 37°F meet. Repeat this step to plot the data for Monday through Saturday. Draw a line to connect all the points.

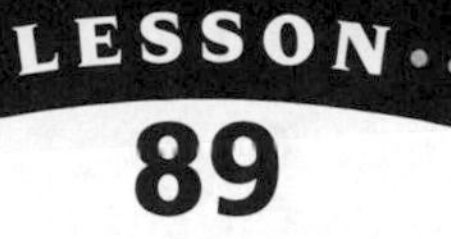

Looking at Line Graphs

What can line graphs tell you?

Line graphs can help you identify patterns that you might not see in a data table. Look at the line graph below. It shows how the temperature in a city changed over one week.

Putting two line graphs side by side helps you see if there is a relationship between the sets of data.

Temperature

Day		Mon.	Tues.	Wed.	Thurs.	Fri.	Sat.	Sun.
Temperature (°F)	80°							
	60°							
	40°							

Temperature and Atmospheric Pressure

Day		Mon.	Tues.	Wed.	Thurs.	Fri.	Sat.	Sun.
Temperature (°F)	80°							
	60°							
	40°							
Air Pressure (millibars)	1040							
	1020							
	1000							

Show What You Know

Compare the two line graphs above to complete this sentence.

As temperature goes ____________________, air pressure goes

____________________.

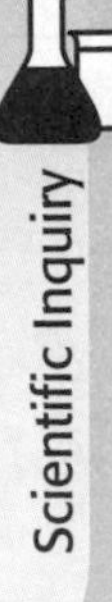

LESSON 90

Drawing Conclusions

What can data tell you?

After you collect your data, you need to **analyze** it. To analyze data means to ask yourself, “What do the data tell me?” Making calculations is often a part of analyzing data. So is making a graph. Graphs can help you see what your data mean.

Data help you draw conclusions. A **conclusion** is an explanation about what the data show. Your conclusion might simply say whether the data support your hypothesis.

An **inference** is also an explanation. But it is based on experience, not on observation. Suppose you put a trashcan out at the curb one night. The next morning, you find the can overturned and the trash scattered across your lawn. You may infer that a wild animal got into the trash. This explanation makes sense. However, this explanation is not a conclusion because you did not see an animal turn over the trashcan.

You can use conclusions and inferences to make predictions. A **prediction** is an idea about what may happen in the future. For example, you might observe that the moon has a regular cycle of phases, or changes in shape. You could use your data to predict when the next phase will happen.

The boy can conclude that a wild animal overturned the trash can.

The boy can infer that a wild animal overturned the trash can

Show What You Know

How is a conclusion different from an inference?

____

____

Fill in the letter to show your answer.

Elevation (m)	Temperature (°C)
0	15.0°
1000	8.5°
2000	2.0°
3000	-4.5°
4000	-11.0°
5000	-17.5°

Make a bar graph of the data. Then use the graph to answer Items 1–2.

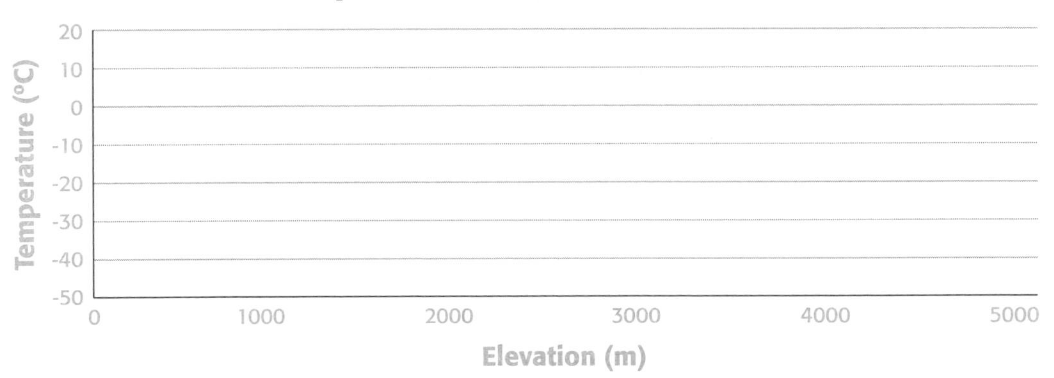

1. **Which pattern does the graph show?**

   Ⓐ As elevation increases, temperature stays the same.

   Ⓑ As elevation increases, temperature increases.

   Ⓒ As elevation decreases, temperature decreases.

   Ⓓ As elevation increases, temperature decreases.

2. **What do you predict the temperature would be at an elevation of 10,000 m?**

   Ⓐ 0.0°C

   Ⓑ -15.0°C

   Ⓒ -25.0°C

   Ⓓ -50.0°C

• • SCIENTIFIC INQUIRY TEST B • •

3. You wake up and look out your window. You see a puddle on the ground. You explain this by stating, "It rained while I was asleep." Your explanation is an example of

Ⓐ a prediction.

Ⓑ a conclusion.

Ⓒ an inference.

Ⓓ an hypothesis.

Short Response

Use the data chart to answer Items 4–5.

A student learns that the density of water is 1 g/cm³. She conducts an experiment to test this hypothesis: "Materials less dense than water float in water." She records her data in the table below.

Material Tested	Density	Floats or Sinks in Water?
Cork	0.24 g/cm³	floats
Corn oil	0.93 g/cm³	floats
Glycerine	1.26 g/cm³	sinks
Aluminum	2.7 g/cm³	sinks

4. What conclusion can the student draw from these data?

____

____

5. How can the student check the results of her experiment?

____

____

LESSON 91

Measuring Tools

What are some measuring tools for science?

Meterstick

Metric ruler

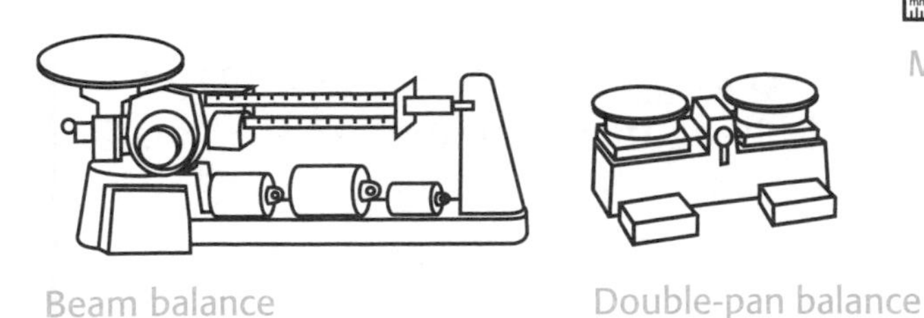

Beam balance

Double-pan balance

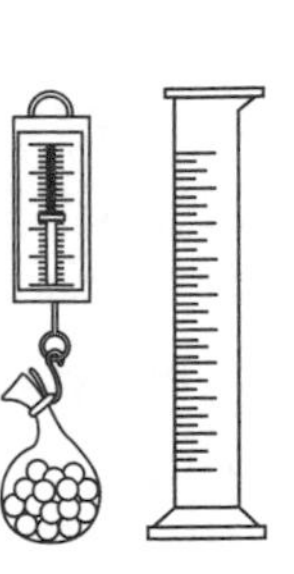

Spring scale, Graduated cylinder, Thermometer

Scientists use tools to make measurements. Distance, or length, is measured with a **metric ruler**, a **meterstick**, or a **tape measure.** A metric ruler measures small distances. A meterstick, which is 1 m long, measures lengths up to a few meters. Distances greater than a few meters are measured with a tape measure.

Mass is measured with a **balance.** A spring scale also can measure mass. But it is more often used to measure force, such as weight. A **graduated cylinder** is used to measure the volume of liquids. A **thermometer** measures temperature.

A **compass** is used to find direction. The compass needle always points toward the North Pole. A **stopwatch** is useful for measuring small lengths of time, such as seconds or fractions of a second. A clock is used to measure longer lengths of time.

Show What You Know

1. Which tool should you use to measure the volume of juice in a juice box? Explain why.

____

____

2. Which tool should you use to measure the width of your desk? Explain why.

____

____

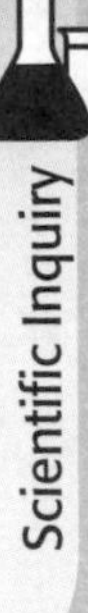
Scientific Inquiry

LESSON 92

Observation Tools

What tools help scientists make observations?

Some things are too small to see using only your eyes. Microscopes help you see small things more clearly. **Microscopes** use lenses to magnify, or make things appear larger.

A **hand lens,** or **magnifier,** is a hand-held microscope with one lens. It magnifies things only a small amount. So, a hand lens is used to look at details you can almost see with your eyes. To see very small objects, like cells, you need a compound microscope. A **compound microscope** uses two lenses to magnify objects.

Objects studied with a compound microscope are placed on slides. The slide sits on the microscope stage. Light passes through the object and into a lens on the nosepiece. This lens magnifies the object. The light then moves up through the tube that holds the eyepiece lens. This lens magnifies the object again.

The nosepiece of a compound microscope has two or three lenses called **objective lenses.** Each lens magnifies an object a different amount. The number of times the lens magnifies is marked near the lens. The magnifying number of the eyepiece is also marked. Multiplying the numbers of the objective and eyepiece lenses tells you how many times larger the object appears.

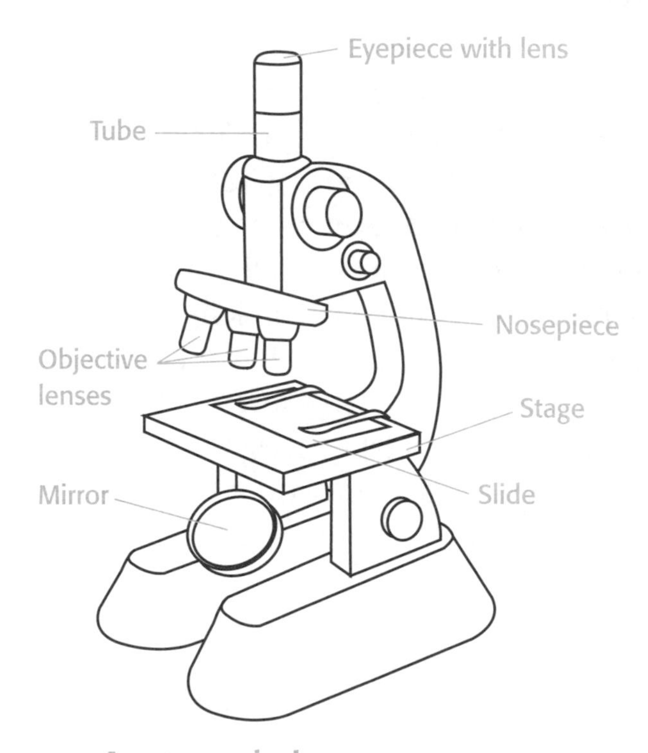

A compound microscope

Show What You Know

If the eyepiece lens of a compound microscope magnifies 10 times and the objective lens also magnifies 10 times, how many times larger does the object you are looking at appear?

____

____

Collecting and Looking at Data

What tools are used to record and analyze data?

Tools like cameras, sound recorders, calculators, and computers help record or analyze data. **Cameras** take still images or snapshots. A video camera also captures images. But these images are moving and show the subject over a period of time. Most video cameras also record sound. A **sound recorder** can also be used to record sounds. Images and sounds captured with tools can be saved for further study.

Data in science are often recorded as numbers. **Calculators** help us analyze, or make sense of, these numbers. A calculator lets you add, subtract, multiply, and divide numbers quickly.

Computers help scientists organize and analyze data. Scientists can use computers to write reports and make charts and graphs. Computers let scientists communicate with other scientists. They are also useful for making calculations and storing data to use later.

Show What You Know

1. When is a calculator useful in science?

____

2. Explain one way a computer is useful in science.

____

LESSON 94

Measuring Mass and Volume

How are mass and volume measured?

Mass is the amount of matter in an object. The main unit of mass is the **gram (g).** Other common units of mass are the **milligram (mg)** and the **kilogram (kg).** One mg is 1/1000 of a g. A kg is 1,000 g. A balance is used to measure mass.

Volume is the amount of space something takes up. For liquids, the **liter (L)** is the main unit of volume. Other common units of volume are the **milliliter (mL)** and the **kiloliter (kL).** One mL is 1/1000 of 1 L. One kL equals 1000 L. The volume of liquids is most often measured with a graduated cylinder.

Solids have volume, too. A solid's shape determines how its volume is measured. A solid, like a box, has a regular shape. Its volume is found by multiplying its length (l), width (w), and height (h). The volume of a solid is usually given in **cubic centimeters (cm^3).**

The volume of a solid that that does not have a regular shape, like a rock, is measured differently. To measure this volume, you use a graduated cylinder as shown in the diagram.

You can change mL into cm^3. One mL of liquid takes up 1 cm^3 of space. So, 1 mL is equal to 1 cm^3.

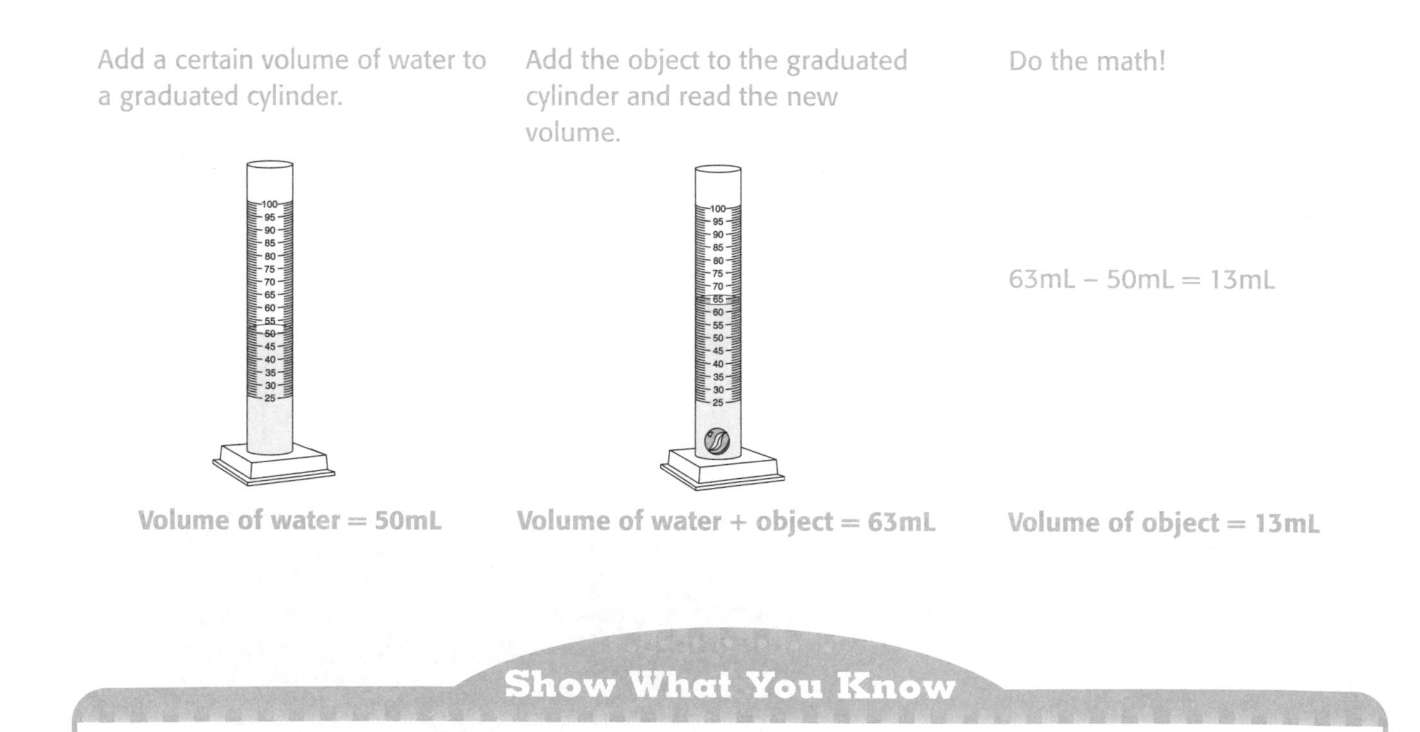

Show What You Know

A box is 3-cm long, 2-cm wide, and 1-cm tall. What is its volume?

____

Measuring Distance and Temperature

How are distance and temperature measured?

In science, length and distance are measured using metric rulers and metersticks. The basic unit of length is the **meter (m).** A meterstick is one meter long.

Long distances are measured in **kilometers (km).** One km equals 1,000 m. Lengths less than 1 m may be measured in **millimeters (mm)** or **centimeters (cm).**

There are 1,000 mm in 1 m, so 1 mm is 1/1,000 of 1 m. Look at the section of a ruler shown in the diagram. Each small line is 1 mm. The longer, numbered lines are centimeters (cm). There are 100 cm in 1 meter, so 1 cm is 1/100 m.

A thermometer is used to measure temperature. There are two commonly used temperature scales—Fahrenheit and Celsius. Common temperatures on both scales are shown.

In science, you use the Celsius temperature scale most often. On this scale, the unit of temperature is the degree **Celsius (°C).** In daily life, most people in the United States measure temperature in degrees **Fahrenheit (°F).**

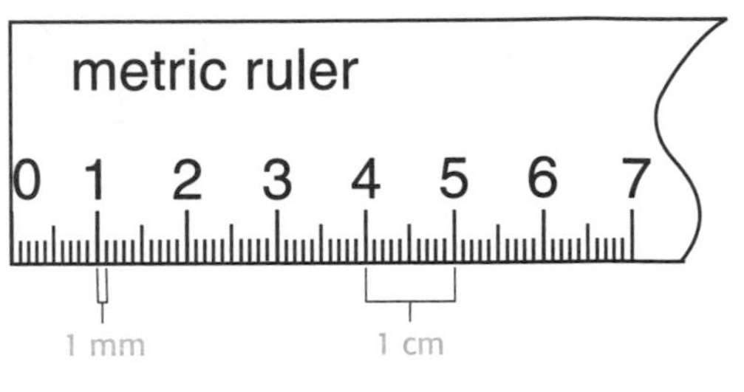

Length measurements

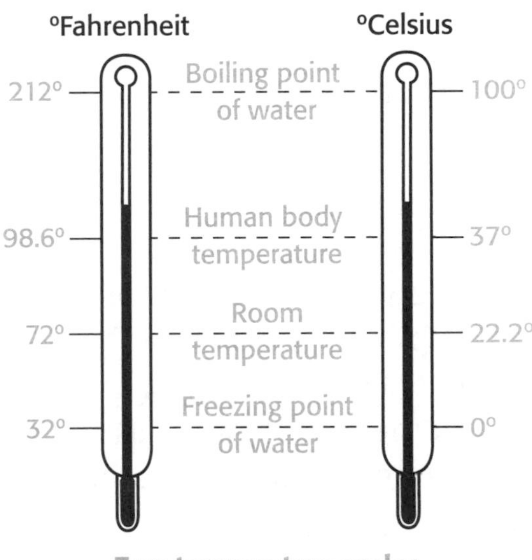

Two temperature scales

Show What You Know

1. What is the boiling point of water in °C and in °F?

____

2. How many meters will you walk if you walk 2 kilometers?

____

LESSON
96

Lab Safety

How can you work safely in the laboratory?

You will work with many materials and tools as you do science. As you work, it is important to keep yourself and others safe. The chart lists rules you should always follow when working in a laboratory. Follow the rules in the chart for all laboratory work. Below are some other rules you should follow, too.

Equipment and Materials Always treat tools and materials with respect. Use tools only as they are supposed to be used. Make sure tools are clean before and after you use them. Take only as much of a material as you need.

Chemical Safety Always handle chemicals carefully. If a spill occurs, tell your teacher immediately. Wear a lab apron and safety goggles when using any chemical. Also wear lab gloves if needed. Label containers before you put something in them. And never use a substance from a container that is not labeled.

Cleanup Always clean up your work area after an investigation. Clean and return tools to their storage area. Ask your teacher what to do with leftover materials. Before leaving the laboratory, wash your hands.

Lab Safety Rules	
☑	Begin work **only** when told to do so by your teacher.
☑	Follow all directions **exactly**.
☑	Ask for help if you don't understand the directions.
☑	Do not eat or drink in the lab and NEVER taste anything.
☑	Keep your work area neat.
☑	Wear a lab apron, safety goggles, and gloves when working with liquids, heat, flames, or chemicals.
☑	Tell your teacher if any accident happens, including spills.

Show What You Know

1. What pieces of safety equipment should you always wear when working with liquids, heat, flames, or chemicals?

____

2. What should you do if an accident occurs in the laboratory?

____

A Multiple Choice

1. **A vet wants to give a dog exactly 375 g of food each day. The vet should**

Ⓐ use a thermometer to find the mass of the food.

Ⓑ use a graduated cylinder to measure the correct amount of food.

Ⓒ use a balance to measure the correct amount of food.

Ⓓ put the dog on a scale each day and weigh it.

2. **Which unit is used to measure how hot boiling water is?**

Ⓐ milliliters

Ⓑ degrees Celsius

Ⓒ cubic centimeters

Ⓓ kilograms

3. **Look at the diagram showing common lab equipment. Which pieces of equipment are used for safety?**

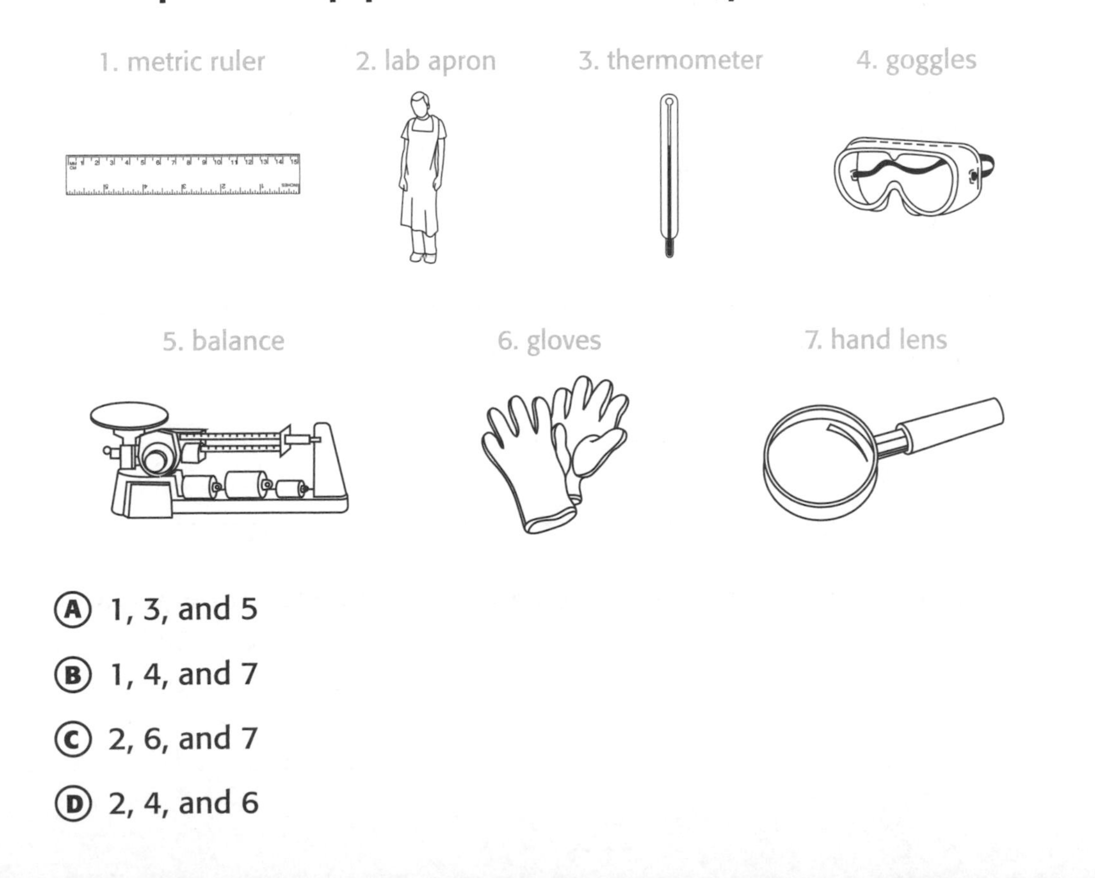

Ⓐ 1, 3, and 5

Ⓑ 1, 4, and 7

Ⓒ 2, 6, and 7

Ⓓ 2, 4, and 6

• • SCIENTIFIC INQUIRY TEST C • •

4. **What tool should you use to measure the length of a paper clip?**

   Ⓐ metric ruler

   Ⓑ meterstick

   Ⓒ tape measure

   Ⓓ balance

5. **Which unit is used to measure the volume of a liquid?**

   Ⓐ degrees Fahrenheit

   Ⓑ kilograms

   Ⓒ centimeters

   Ⓓ milliliters

Short Response

Use the illustration to answer Items 6–7.

A student is using chemicals for an investigation. She has put 100 mL of a chemical in a graduated cylinder. Now she is pouring the chemical into a beaker as shown.

6. **What should the student have done to the beaker before pouring liquid into it?**

   ____

   ____

7. **What is wrong with the safety equipment the student is wearing?**

   ____

   ____

Check your answers for Science Inquiry Test A on pages 127-128.

Multiple Choice

1. **Look at the experimental setup below. Which of the following is most likely the hypothesis for this experiment?**

   (A) Saltwater has a lower boiling point than freshwater.

   Since a thermometer and hot plate are shown, the experiment is likely testing the boiling point of saltwater.

2. **A student plans an investigation about plant growth. The procedure has four trials. The type of plant, type of soil, and amount of water remain the same in each trial. Each group of plants receives different amounts of light. The independent variable in this experiment is the**

   (B) amount of light.

   The variable being tested is the amount of light each group of plants receives.

3. **Dylan does an experiment. The data he gathers do not support his hypothesis. Dylan should**

   (C) report his results exactly as he observed them.

   Reporting the results as they are observed keeps data honest.

Short Response

4. **Which planets are farther from the sun than Earth is?**

   Mars, Jupiter, Saturn, Uranus, Neptune, Pluto

   The distance of each of these six planets from the sun is greater than 149,598,000 miles—the distance between Earth and the sun.

5. **Which planet can orbit the sun four times in the amount of time it takes Earth to orbit the sun once?**

   Mercury

   Mercury orbits the sun in 0.24 Earth years, or about one-fourth the time it takes Earth to orbit the sun. So it can orbit the sun four times in the time it takes Earth to orbit the sun once.

6. **Compare the distance of the planets from the sun to the time it takes each planet to orbit the sun. What pattern can you see?**

   As the distance of a planet from the sun increases, the amount of time it takes that planet to orbit the sun also increases.

   The farther a planet is from the sun, the longer it takes to travel once around it.

Check your answers for Science Inquiry Test B on pages 133-134.

Multiple Choice

1. Which pattern does the graph show?

D As elevation increases, temperature decreases.

The sloping line shows that as elevation went up, temperature went down.

2. What do you predict the temperature would be at an elevation of 10,000 m?

D -50.0°C

The graph shows that temperature decreases as elevation increases. So, the temperature at 10,000 meters would be lower than the temperature at 5,000 meters. The pattern shown by the line makes it clear that the temperature would be closer to -50°C than -25°C.

3. You wake up and look out your window. You see a puddle on the ground. You explain this by stating, "It rained while I was asleep." Your explanation is an example of

C an inference.

You did not observe it raining, so your explanation is an inference—an explanation based on experience but not observation.

4. What conclusion can the student draw from these data?

Material Tested	Density	Floats or Sinks in Water?
Cork	0.24 g/cm³	floats
Corn oil	0.93 g/cm³	floats
Glycerine	1.26 g/cm³	sinks
Aluminum	2.7 g/cm³	sinks

The data support the hypothesis.

The data show that only the materials with densities less than the density of water float in the water. So the data support the hypothesis.

5. How can the student check the results of her experiment?

Repeat the experiment to see if the results are the same.

Repeating experiments improves the chances that the results will be correct.

Check your answers for Scientific Inquiry Test C on pages 141-142.

A Multiple Choice

1. **A vet wants to give a dog exactly 375 g of food each day. The vet should**

   C use a balance to measure the correct amount of food.

   Grams are a unit of mass. Mass is measured with a balance.

2. **Which unit is used to measure how hot boiling water is?**

   B degrees Celsius

   In science, temperature is measured in degrees Celsius.

3. **Look at the diagram showing common lab equipment. Which pieces of equipment are used for safety?**

   D 2, 4, and 6

   A lab apron, goggles, and gloves are all pieces of safety equipment.

4. **What tool should you use to measure the length of a paper clip?**

   A metric ruler

   Metric rulers are used to measure length. A paper clip is small, so a metric ruler is a better choice than a meterstick.

5. **Which unit is used to measure the volume of a liquid?**

   D milliliters

   The volume of liquids is usually measured in liters or milliliters.

Short Response

6. **What should the student have done to the beaker before pouring liquid into it?**

   The beaker should have been labeled before a chemical was poured into it.

   Always label a container before adding something to it. And never use a substance from an unlabeled container.

7. **What is wrong with the safety equipment the student is wearing?**

   The student is not using her safety goggles. They should be covering her eyes.

   Safety goggles should always be worn when working with chemicals.

• • PRACTICE TEST • •

You will find two kinds of questions on the Practice Test—multiple-choice and short-response. Read these examples.

Multiple Choice

Fill in the letter to show your answer.

1. All of the populations living within an area make up an

Ⓐ organism

Ⓑ species

Ⓒ community

Ⓓ ecosystem

Together, all of the populations living in an area make up a community. The correct answer is C, so you should fill in the letter C to show your answer.

Strategies and Tips

As you complete the Practice Test, remember to:

- read each question carefully.
- read all of the answer choices before deciding which one is correct.
- study pictures, charts, and diagrams carefully before answering.
- skip difficult questions. Come back to them later when you have more time.
- write clear and complete short responses.

Short Response

Use the picture to answer Item 2.

2. What roles do each of the organisms in the picture have in an ecosystem?

The diagram shows two organisms—a caterpillar and a leaf. Your answer should explain the purpose each organism has in its environment. Here is an example of a good short response.

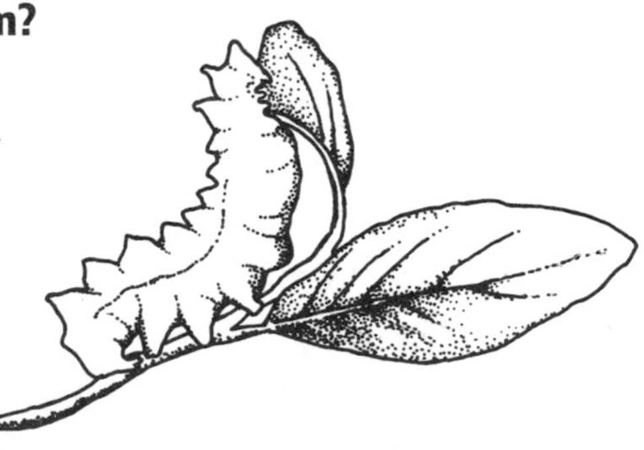

The leaf belongs to a plant. A plant is a producer. It uses water, carbon dioxide, and the energy in sunlight to make its own food. The caterpillar is a consumer. It eats plants to stay alive. The caterpillar and leaf are part of a food chain in an ecosystem.

Multiple Choice

Fill in the letter to show your answer.

1. **Frogs, snakes, and fish are alike because they all**
   - Ⓐ live in water.
   - Ⓑ have backbones.
   - Ⓒ breathe with lungs.
   - Ⓓ have scales.

2. **Most fungi are decomposers. Fungi are important to food chains because they**
   - Ⓐ make food by photosynthesis.
   - Ⓑ feed on producers.
   - Ⓒ return nutrients to the soil.
   - Ⓓ are hosts to parasites.

3. **Raj uses a screwdriver to pry the lid off a paint can. What type of simple machine did Raj use?**
   - Ⓐ inclined plane
   - Ⓑ wheel and axle
   - Ⓒ lever
   - Ⓓ wedge

4. **Which of the following is an example of a chemical change?**
   - Ⓐ rusting of iron
   - Ⓑ freezing of water
   - Ⓒ dissolving of sugar
   - Ⓓ tearing of paper

• • PRACTICE TEST • •

Use the picture to answer Items 5–7.

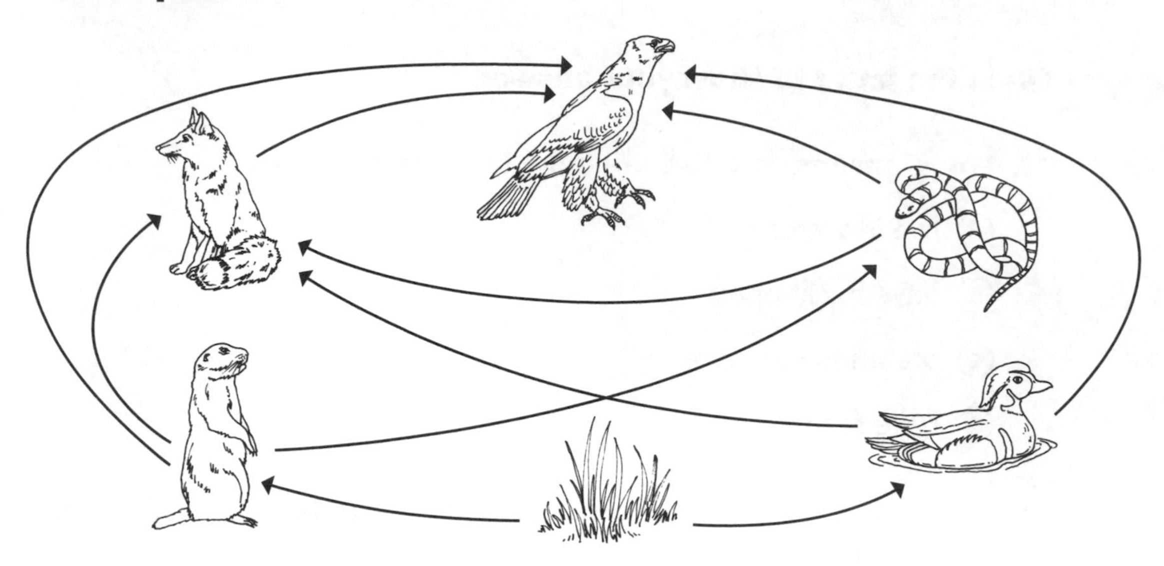

5. Which populations would most likely increase in size if snakes were removed from this food web?

Ⓐ grasses and ducks

Ⓑ prairie dogs and ducks

Ⓒ prairie dogs and red foxes

Ⓓ golden eagles and ducks

6. In this food web, grasses are

Ⓐ predators.

Ⓑ decomposers.

Ⓒ consumers.

Ⓓ producers.

7. Which group shows a correct and complete path of energy flow in this food web?

Ⓐ grass, snake, red fox

Ⓑ grass, prairie dog, red fox, golden eagle

Ⓒ grass, duck, prairie dog

Ⓓ golden eagle, snake, duck, grass

8. If you want to find out how the mass of a puppy changes each week, you should

Ⓐ measure the puppy's length each day with a ruler.

Ⓑ feed the puppy 100 g of food each day.

Ⓒ use a balance to measure the mass of the puppy each week.

Ⓓ record the puppy's growth with a video camera.

9. Of the following, the planet that makes one complete trip around the sun in the shortest amount of time is

Ⓐ Mars.

Ⓑ Mercury.

Ⓒ Jupiter.

Ⓓ Neptune.

10. You move a magnet toward an object. The object begins moving toward the magnet. The object is most likely

Ⓐ made of plastic.

Ⓑ made of wood.

Ⓒ made of aluminum.

Ⓓ made of iron.

11. What energy changes occur when wood burns?

Ⓐ Chemical energy changes to light and heat.

Ⓑ Heat energy changes to mechanical and light energy.

Ⓒ Light and heat change to chemical energy.

Ⓓ Chemical energy changes to light and electrical energy.

• • PRACTICE TEST • •

12. Predict what will happen to the plant if the pot is stood upright.

Ⓐ The plant's stem will stop growing.

Ⓑ The plant's roots will stop growing and the plant will die.

Ⓒ The plant's roots will bend to grow downward and its stem will bend to grow upward.

Ⓓ The plant's stem and leaves will bend to grow away from sunlight.

13. A tangerine looks orange because

Ⓐ it absorbs only orange light and reflects light of all other colors.

Ⓑ it refracts only orange light.

Ⓒ it reflects only orange light, and absorbs light of all other colors.

Ⓓ orange light waves have the longest wavelengths.

14. Which piece of lab equipment is *not* used for safety?

Ⓐ gloves

Ⓑ laboratory coat

Ⓒ goggles

Ⓓ thermometer

15. The spoon in this glass appears bent or broken because

Ⓐ the speed of light changes as it moves from one material into another.

Ⓑ the spoon is unable to reflect light in water.

Ⓒ light waves travel in straight lines.

Ⓓ the spoon absorbs more light waves than the water.

16. How do plants cause the weathering of rocks?

Ⓐ Leaves shade the rocks, protecting them from the sun.

Ⓑ Roots grow in cracks in rocks, splitting the rocks apart.

Ⓒ Plants steal minerals from the rocks.

Ⓓ As plants die, they add minerals to the rocks.

17. The shape of this bird's beak is best adapted for

Ⓐ cracking hard seeds.

Ⓑ scooping small fish from the water.

Ⓒ reaching deep into flowers to feed on nectar.

Ⓓ digging insects from the bark of trees.

18. The length of time it takes Earth to make one complete path around the sun is

Ⓐ 1 day.

Ⓑ 1 week.

Ⓒ 1 month.

Ⓓ 1 year.

• • PRACTICE TEST • •

Use this information to answer Items 19–21.

Three students sat quietly for 5 minutes. Each student then measured his or her pulse rate. The students then spent 5 minutes running in place and measured their pulse rates again. The results are given below.

Pulse Rates Before and After Exercise

Student	Pulse Rate at Rest (Beats/Minute)	Pulse Rate After Exercise (Beats/Minute)
1	70	92
2	75	102
3	74	98

19. What hypothesis was tested by this experiment?

Ⓐ Pulse rate is measured in beats per minute.

Ⓑ Sleeping has no effect on pulse rate.

Ⓒ Exercise has no effect on breathing rate or pulse rate.

Ⓓ Pulse rate increases with exercise.

20. Which of the following conclusions do the data support?

Ⓐ Pulse rate is not affected by exercise.

Ⓑ Exercise causes pulse rate to increase.

Ⓒ Exercise causes pulse rate to decrease.

Ⓓ Running in place for 5 minutes is good for the heart.

21. What is the independent variable in this experiment?

Ⓐ exercise

Ⓑ rest

Ⓒ pulse rate

Ⓓ time

22. What change might pollution be causing to the ozone layer over the north and south poles?

Ⓐ The ozone layer is becoming thicker.

Ⓑ Holes are forming in the ozone layer.

Ⓒ Scientists have observed no changes to the ozone layer.

Ⓓ The ozone layer blocks more of the sun's harmful rays.

23. Which of the following is *not* directly needed for an animal's survival?

Ⓐ soil

Ⓑ food

Ⓒ oxygen

Ⓓ water

24. Which of the following is a renewable energy resource?

Ⓐ oil

Ⓑ coal

Ⓒ wind

Ⓓ natural gas

25. Which of the following would you use to find the volume of a liquid?

Ⓐ graduated cylinder

Ⓑ thermometer

Ⓒ balance

Ⓓ spring scale

26. Which pair of words below best matches the word pairs in the box?

BUTTERFLY – LARVA
TREE – SEEDLING

Ⓐ WORM – INVERTEBRATE

Ⓑ EGG – MAMMAL

Ⓒ CHICKEN – FEATHER

Ⓓ FROG – TADPOLE

27. Which will tell you the most about the volume of an object?

Ⓐ lifting it

Ⓑ smelling it

Ⓒ looking at it

Ⓓ taking it apart

28. A dolphin sends out sound waves. Some of these waves return to the dolphin as an echo. From these waves, the dolphin knows what kinds of objects are in front of it. The property of sound that produces an echo is

Ⓐ transmission.

Ⓑ reflection.

Ⓒ refraction.

Ⓓ absorption.

The data table and graph show how temperature affected seed germination. Use both to answer Items 29–31.

Seed Germination and Temperature

Temperature °C	Days Needed to Sprout
13°	22
15°	18
18°	15
21°	12
24°	9

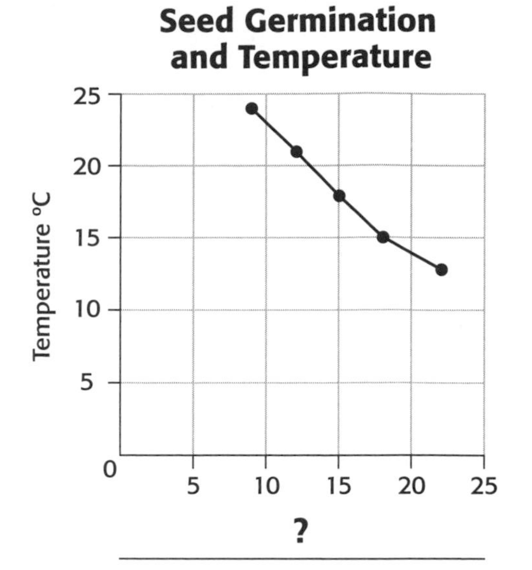

29. Based on the table, how long will it take for the same kind of seed to sprout at 11°C?

Ⓐ more than 22 days

Ⓑ between 18 and 22 days

Ⓒ between 12 and 18 days

Ⓓ less than 9 days

30. A good label for the graph's horizontal axis is

Ⓐ Number of Days

Ⓑ Degrees Celsius

Ⓒ Number of Sprouts

Ⓓ Seed Germination

31. Which statement does the graph show?

Ⓐ The larger the seed, the shorter the time needed to sprout.

Ⓑ Temperature does not affect how long it takes seeds to sprout.

Ⓒ Seeds sprout more quickly at lower temperatures.

Ⓓ Seeds sprout more quickly at warmer temperatures.

32. Each picture shows a circuit. In which circuit or circuits does electricity have only one path to follow?

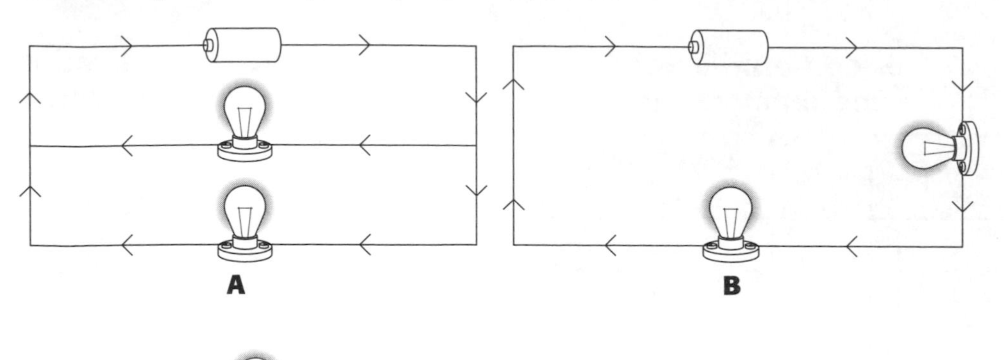

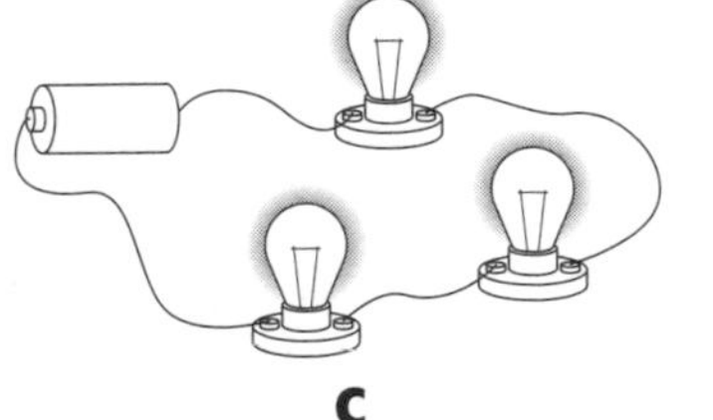

Ⓐ circuit A only

Ⓑ circuits A and B

Ⓒ circuits B and C

Ⓓ circuits A and C

33. Water changes to ice when

Ⓐ heat energy is added.

Ⓑ heat energy is taken away.

Ⓒ a chemical change takes place.

Ⓓ it condenses.

34. Juan does an experiment. The data he gathers do not seem to support his hypothesis. What should he do?

Ⓐ Change his hypothesis to match the data.

Ⓑ Change his data to agree with his hypothesis.

Ⓒ Copy data from a classmate who got different results.

Ⓓ Report his data exactly as he observed them.

Record your answers in the space provided below each question.

Use the pictures to answer Item 35.

35. **Look at the two pictures. In which scene will the boy be able to bring the bicycle to a stop more quickly? Why?**

____

____

____

____

____

____

• • PRACTICE TEST • •

Use the illustration to answer Items 36–37.

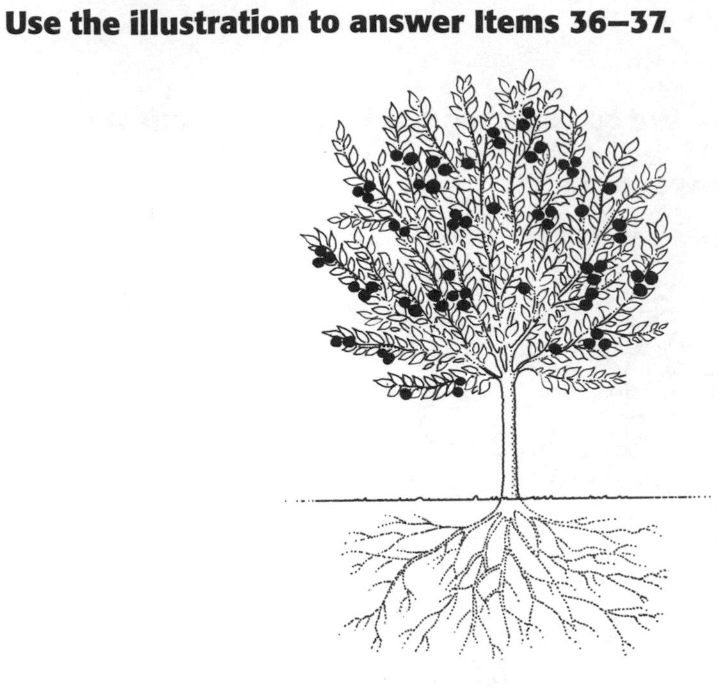

36. In the first column of the chart below, identify *three* structures that help this tree survive. In the second column, explain how the tree uses each structure to help it survive.

Tree Structure	How the Tree Uses the Structure to Help It Survive

37. What parts of the tree are used for reproduction?

___

___

38. Wolves, like many animals, have adaptations that are learned and adaptations they inherit from their parents. Place a check (✔) in the correct column of the chart below to identify whether each adaptation is learned or inherited in wolves.

Adaptation	Learned	Inherited
Thick fur		
Hunting ability		
Sharp teeth		

39. A plant's stem and leaves grow toward sunlight. This is an inherited adaptation. How does it help a plant survive?

____

____

____

____

____

40. Hibernation is an inherited adaptation. How does hibernation help an animal survive winter?

____

____

____

____

____

• • PRACTICE TEST • •

Use the diagram to answer Items 41-43.

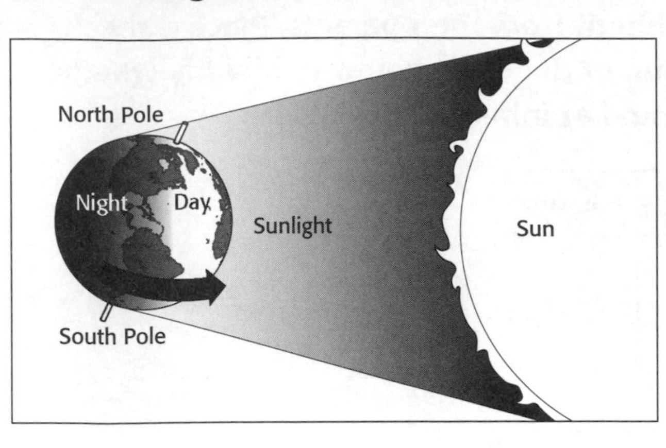

41. What is the name for the movement of Earth shown in the diagram?

____

42. How long does it take for this movement to occur once?

____

43. Use the diagram to help you explain what causes day and night on Earth.

____

____

____

____

____

____